# Plant Pathology

# Plant Pathology

Edited by
**Tomos Webb**

Larsen & Keller
www.larsen-keller.com

Plant Pathology
Edited by Tomos Webb
ISBN: 978-1-63549-224-8 (Hardback)

**⊟ Larsen & Keller**

Published by Larsen and Keller Education,
5 Penn Plaza,
19th Floor,
New York, NY 10001, USA

**Cataloging-in-Publication Data**

Plant pathology / edited by Tomos Webb.
    p. cm.
Includes bibliographical references and index.
ISBN 978-1-63549-224-8
1. Plant diseases. 2. Diseased plants. 3. Agricultural pests. 4. Phytopathogenic microorganisms--Control.
I. Webb, Tomos.
SB601 .P53 2017
632--dc23

The publisher's policy is to use permanent paper from mills that operate a sustainable forestry policy. Furthermore, the publisher ensures that the text paper and cover boards used have met acceptable environmental accreditation standards.

Printed and bound in the United States of America.

For more information regarding Larsen and Keller Education and its products, please visit the publisher's website www.larsen-keller.com

# Table of Contents

# Preface

This book provides extensive details about the vast topic of plant pathology, also known as phytopathology. It deals with the study of diseases in the plants caused by pathogens and other physical factors. Plant pathology includes topics like pathogen identification, plant disease epidemiology, disease etiology, etc. The pathogens responsible for these diseases are bacteria, virus, protozoa, parasitic plants, etc. This book is a valuable compilation of various topics, ranging from the most basic to the most complex ones. For those who have an interest in plant pathology, this book will prove to be a valuable source of knowledge.

To facilitate a deeper understanding of the contents of this book a short introduction of every chapter is written below:

Chapter 1- The scientific study of diseases in plants caused either by pathogens or by environmental conditions is known as plant pathology. The organisms that cause these infections in plants are fungi, bacteria, viruses, nematodes and parasitic plants. This chapter is an overview of the subject matter incorporating all the major aspects of plant pathology.

Chapter 2- Pathogens are what produce diseases. Generally, the term is used to describe an infectious agent such as virus, bacterium or a fungus. This chapter focuses on oomycete and the significance of oomycete plant pathogens and also focuses on bacteria and bacterial plant pathogens. The topics discussed in the chapter are of great importance to broaden the existing knowledge on plant pathology.

Chapter 3- This chapter provides the reader with a broad understanding on the various disorders found in plants. It particularly focuses on physiological plant disorder, which is dissimilar to plant diseases caused by pathogens. Altering environmental conditions can usually prevent physiological plant disorder. The major causes of physiological plant disorder discussed in this chapter are phytotoxicity, boron deficiency, calcium deficiency, magnesium deficiency and phosphorus deficiency.

Chapter 4- This chapter will provide an integrated understanding of plant diseases. It offers an insightful focus, keeping in mind the complex subject matter. Organisms of various kinds cause these disorders. Phyllody, citrus canker, powdery scab, wilt disease, top dying disease and barley yellow dwarf are broadly explained in this chapter.

Chapter 5- The chapter explicates diseases such as glomerella graminicola, common bunt, spot blotch, wheat leaf rust, and leaf rust. Spot blotch is a leaf disease of wheat caused by Cochliobolus sativus while leaf rust is a fungal disease of barley caused by Puccinia hordei. Also known as brown rust, it is the most serious rust disease on barley. The aspects elucidated in this chapter are of vital importance, and provide a better understanding of plant pathology.

Chapter 6- Plant disease epidemiology is the study of disease in plant pollutions. Plant diseases occur due to pathogens such as bacteria, viruses, fungi and parasitic plants. Early detection and accurate diagnosis is essential for effective management of plant disease. This chapter provides a comprehensive overview of plant disease.

Chapter 7- Protecting plants from diseases is very important and this chapter explains the various protections of plants from diseases. Plant disease resistance, diseases resistance in fruit and vegetables, pest control and fungicide are some of the protections, which are discussed in the following content.

Chapter 8- Thigmonasty is the response of a plant or fungus to touch or vibration, while the ability of reducing the negative fitness effects caused by herbivory is known as the plant tolerance to herbivory. This chapter elaborates on other defense mechanisms also, which include, raphide, inducible plan defenses against herbivory and plant use of endophytic fungi in defense.

Chapter 9- The chemicals that regulate plant growth are known as plant hormones. Plant hormones occur within the plant and occur in extremely low concentrations. Hormones also determine the formation of flowers, stems, leaves and the shedding of leaves. The following content also focuses on forest pathology, which is the research of both biotic and abiotic maladies affecting the health of a forest ecosystem. Forest pathology is a part of the comprehensive approach of forest protection.

I would like to share the credit of this book with my editorial team who worked tirelessly on this book. I owe the completion of this book to the never-ending support of my family, who supported me throughout the project.

**Editor**

# Introduction to Plant Pathology

The scientific study of diseases in plants caused either by pathogens or by environmental conditions is known as plant pathology. The organisms that cause these infections in plants are fungi, bacteria, viruses, nematodes and parasitic plants. This chapter is an overview of the subject matter incorporating all the major aspects of plant pathology.

## Plant Pathology

Plant pathology (also phytopathology) is the scientific study of diseases in plants caused by pathogens (infectious organisms) and environmental conditions (physiological factors). Organisms that cause infectious disease include fungi, oomycetes, bacteria, viruses, viroids, virus-like organisms, phytoplasmas, protozoa, nematodes and parasitic plants. Not included are ectoparasites like insects, mites, vertebrate, or other pests that affect plant health by consumption of plant tissues. Plant pathology also involves the study of pathogen identification, disease etiology, disease cycles, economic impact, plant disease epidemiology, plant disease resistance, how plant diseases affect humans and animals, pathosystem genetics, and management of plant diseases.

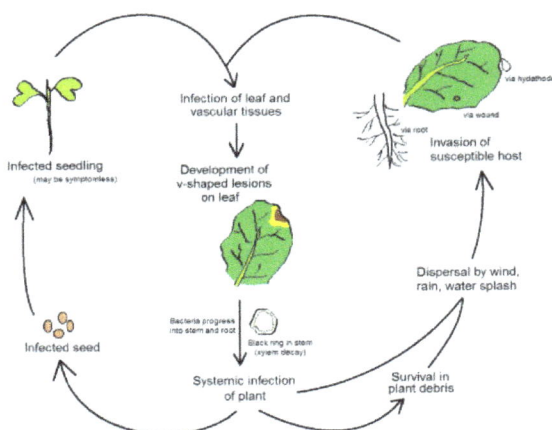

Life cycle of the black rot pathogen, *Xanthomonas campestris* pathovar *campes*

### Overview

Control of plant diseases is crucial to the reliable production of food, and it provides significant reductions in agricultural use of land, water, fuel and other inputs. Plants in both natural and cultivated populations carry inherent disease resistance, but there are numerous examples of devastating plant disease impacts, as well as recurrent severe plant diseases. However, disease control is reasonably successful for most crops.

Disease control is achieved by use of plants that have been bred for good resistance to many diseases, and by plant cultivation approaches such as crop rotation, use of pathogen-free seed, appropriate planting date and plant density, control of field moisture, and pesticide use. Across large regions and many crop species, it is estimated that dis-eases typically reduce plant yields by 10% every year in more developed settings, but yield loss to diseases often exceeds 20% in less developed settings. Continuing advances in the science of plant pathology are needed to improve disease control, and to keep up with changes in disease pressure caused by the ongoing evolution and movement of plant pathogens and by changes in agricultural practices. Plant diseases cause major economic losses for farmers worldwide. The Food and Agriculture Organization estimates indeed that pests and diseases are responsible for about 25% of crop loss. To solve this issue, new methods are needed to detect diseases and pests early, such as novel sensors that detect plant odours and spectroscopy and biophotonics that are able to diagnostic plant health and metabolism.

## Plant Pathogens

Powdery mildew, a biotrophic fungus

## Fungi

Most phytopathogenic fungi belong to the Ascomycetes and the Basidiomycetes.

The fungi reproduce both sexually and asexually via the production of spores and other structures. Spores may be spread long distances by air or water, or they may be soilborne. Many soil inhabiting fungi are capable of living saprotrophically, carrying out the part of their life cycle in the soil. These are known as facultative saprotrophs.

Fungal diseases may be controlled through the use of fungicides and other agriculture practices. However, new races of fungi often evolve that are resistant to various fungicides.

Biotrophic fungal pathogens colonize living plant tissue and obtain nutrients from living host cells. Necrotrophic fungal pathogens infect and kill host tissue and extract nutrients from the dead host cells. See the powdery mildew and rice blast images, below.

Rice blast, caused by a necrotrophic fungus

Significant fungal plant pathogens include:

## Ascomycetes

- *Fusarium* spp. (causal agents of Fusarium wilt disease)
- *Thielaviopsis* spp. (causal agents of: canker rot, black root rot, *Thielaviopsis* root rot)
- *Verticillium* spp.
- *Magnaporthe grisea* (causal agent of rice blast)
- *Sclerotinia sclerotiorum* (causal agent of cottony rot)

## Basidiomycetes

- *Ustilago* spp. (causal agents of smut)
- *Rhizoctonia* spp.
- *Phakospora pachyrhizi* (causal agent of soybean rust)
- *Puccinia* spp. (causal agents of severe rusts of virtually all cereal grains and cultivated grasses)
- *Armillaria* spp. (the so-called honey fungus species, which are virulent pathogens of trees and produce edible mushrooms)

## Fungus-like Organisms

## Oomycetes

The oomycetes are not true fungi but are fungus-like organisms. They include some of the most destructive plant pathogens including the genus *Phytophthora*, which includes the causal agents of potato late blight and sudden oak death. Particular species of oomycetes are responsible for root rot.

Despite not being closely related to the fungi, the oomycetes have developed very similar infection strategies. Oomycetes are capable of using effector proteins to turn off a plant's defenses in its infection process. Plant pathologists commonly group them with fungal pathogens.

Significant oomycete plant pathogens

- *Pythium* spp.
- *Phytophthora* spp., including the causal agent of the Great Irish Famine (1845–1849)

## Phytomyxea

Some slime molds in Phytomyxea cause important diseases, including club root in cabbage and its relatives and powdery scab in potatoes. These are caused by species of *Plasmodiophora* and *Spongospora*, respectively.

## Bacteria

Crown gall disease caused by Agrobacterium

Most bacteria that are associated with plants are actually saprotrophic and do no harm to the plant itself. However, a small number, around 100 known species, are able to cause disease. Bacterial diseases are much more prevalent in subtropical and tropical regions of the world.

Most plant pathogenic bacteria are rod-shaped (bacilli). In order to be able to colonize the plant they have specific pathogenicity factors. Five main types of bacterial pathogenicity factors are known: uses of cell wall–degrading enzymes, toxins, effector proteins, phytohormones and exopolysaccharides.

Pathogens such as *Erwinia* species use cell wall–degrading enzymes to cause soft rot. *Agrobacterium* species change the level of auxins to cause tumours with phytohormones. Exopolysaccharides are produced by bacteria and block xylem vessels, often leading to the death of the plant.

Bacteria control the production of pathogenicity factors via quorum sensing.

Vitis vinifera with "Ca. Phytoplasma vitis" infection

Significant bacterial plant pathogens:

- Burkholderia

- Proteobacteria

    o *Xanthomonas* spp.

    o *Pseudomonas* spp.

- Pseudomonas syringae pv. tomato causes tomato plants to produce less fruit, and it "continues to adapt to the tomato by minimizing its recognition by the tomato immune system."

Phytoplasmas ('Mycoplasma-like organisms') and spiroplasmas

*phytoplasma*

*Phytoplasma* and *Spiroplasma* are a genre of bacteria that lack cell walls and are related to the mycoplasmas, which are human pathogens. Together they are referred to as the mollicutes. They also tend to have smaller genomes than most other bacteria. They are normally transmitted by sap-sucking insects, being transferred into the plants phloem where it reproduces.

Tobacco mosaic virus

## Viruses, Viroids and Virus-like Organisms

There are many types of plant virus, and some are even asymptomatic. Under normal circumstances, plant viruses cause only a loss of crop yield. Therefore, it is not economically viable to try to control them, the exception being when they infect perennial species, such as fruit trees.

Most plant viruses have small, single-stranded RNA genomes. However some plant viruses also have double stranded RNA or single or double stranded DNA genomes. These genomes may encode only three or four proteins: a replicase, a coat protein, a movement protein, in order to allow cell to cell movement through plasmodesmata, and sometimes a protein that allows transmission by a vector. Plant viruses can have several more proteins and employ many different molecular translation methods.

Plant viruses are generally transmitted from plant to plant by a vector, but mechanical and seed transmission also occur. Vector transmission is often by an insect (for example, aphids), but some fungi, nematodes, and protozoa have been shown to be viral vectors. In many cases, the insect and virus are specific for virus transmission such as the beet leafhopper that transmits the curly top virus causing disease in several crop plants.

## Nematodes

Root-knot nematode galls

Nematodes are small, multicellular wormlike animals. Many live freely in the soil, but there are some species that parasitize plant roots. They are a problem in tropical and subtropical regions of the world, where they may infect crops. Potato cyst nematodes (*Globodera pallida* and *G. rostochiensis*) are widely distributed in Europe and North and South America and cause $300 million worth of damage in Europe every year. Root knot nematodes have quite a large host range, whereas cyst nematodes tend to be able to infect only a few species. Nematodes are able to cause radical changes in root cells in order to facilitate their lifestyle.

## Protozoa and Algae

There are a few examples of plant diseases caused by protozoa (e.g., *Phytomonas*, a kinetoplastid). They are transmitted as zoospores that are very durable, and may be able to survive in a resting state in the soil for many years. They have also been shown to transmit plant viruses.

When the motile zoospores come into contact with a root hair they produce a plasmodium and invade the roots.

Some colourless parasitic algae (e.g., *Cephaleuros*) also cause plant diseases.

## Parasitic Plants

Parasitic plants such as mistletoe and dodder are included in the study of phytopathology. Dodder, for example, is used as a conduit either for the transmission of viruses or virus-like agents from a host plant to a plant that is not typically a host or for an agent that is not graft-transmissible.

## Common Pathogenic Infection Methods

- Cell wall-degrading enzymes: These are used to break down the plant cell wall in order to release the nutrients inside.

- Toxins: These can be non-host-specific, which damage all plants, or host-specific, which cause damage only on a host plant.

- Effector proteins: These can be secreted into the extracellular environment or directly into the host cell, often via the Type three secretion system. Some effectors are known to suppress host defense processes. This can include: reducing the plants internal signaling mechanisms or reduction of phytochemicals production. Bacteria, fungus and oomycetes are known for this function.

## Physiological Plant Disorders

Significant abiotic disorders can be caused by:

### *Natural*

Drought

Frost damage and breakage by snow and hail

Flooding and poor drainage

Nutrient deficiency

Salt deposition and other soluble mineral excesses (e.g., gypsum)

Wind (windburn and breakage by hurricanes and tornadoes)

Lightning and wildfire (also often man-made)

***Man-made*** (arguably not abiotic, but usually regarded as such)

Soil compaction

Pollution of air, soil, or both

Salt from winter road salt application or irrigation

Herbicide over-application

Poor education and training of people working with plants (e.g. lawnmower damage to trees)

Vandalism

Orchid leaves with viral infections

# Epidemiology

# Disease Resistance

# Management

Quarantine

A diseased patch of vegetation or individual plants can be isolated from other, healthy growth. Specimens may be destroyed or relocated into a greenhouse for treatment or study. Another option is to avoid the introduction of harmful nonnative organisms by controlling all human traffic and activity (e.g., AQIS), although legislation and enforcement are crucial in order to ensure lasting effectiveness.

Cultural

Farming in some societies is kept on a small scale, tended by peoples whose culture includes farming traditions going back to ancient times. (An example of such traditions would be lifelong training in techniques of plot terracing, weather anticipation and response, fertilization, grafting, seed care, and dedicated gardening.) Plants that are intently monitored often benefit from not only active external protection but also a greater overall vigor. While primitive in the sense of being the most labor-intensive solution by far, where practical or necessary it is more than adequate.

Plant resistance

Sophisticated agricultural developments now allow growers to choose from among systematically cross-bred species to ensure the greatest hardiness in their crops, as suited for a particular region's pathological profile. Breeding practices have been perfected over centuries, but with the advent of genetic manipulation even finer control of a crop's immunity traits is possible. The engineering of food plants may be less rewarding, however, as higher output is frequently offset by popular suspicion and negative opinion about this "tampering" with nature.

## Chemical

Many natural and synthetic compounds can be employed to combat the above threats. This method works by directly eliminating disease-causing organisms or curbing their spread; however, it has been shown to have too broad an effect, typically, to be good for the local ecosystem. From an economic standpoint, all but the simplest natural additives may disqualify a product from "organic" status, potentially reducing the value of the yield.

## Biological

Crop rotation may be an effective means to prevent a parasitic population from becoming well-established, as an organism affecting leaves would be starved when the leafy crop is replaced by a tuberous type, etc. Other means to undermine parasites without attacking them directly may exist.

## Integrated

The use of two or more of these methods in combination offers a higher chance of effectiveness.

# Timeline of Plant Pathology

300–286 BC Theophrastus, father of botany, wrote and studied diseases of trees, cereals and legumes

1665 Robert Hooke illustrates a plant-pathogenic fungal disease, rose rust

1675 Anton van Leeuwenhouek invents the compound microscope, in 1683 describes bacteria seen with the microscope

1729 Pier Antonio Micheli, father of mycology, observes spores for the first time, conducts germination experiments

1755 Tillet reports on treatment of seeds

1802 Lime sulfur first used to control plant disease

1845–1849 Potato late blight epidemic in Ireland

1853 Heinrich Anton de Bary father of modern mycology, establishes that fungi are the cause, not the result, of plant diseases, publishes "Untersuchungen uber die Brandpilze"

1858 Julius Kühn publishes "Die Krankheiten der Kultergewachse"

1865 M. Planchon discovers a new species of *Phylloxera*, which was named *Phylloxera vastatrix.*

1868–1882 Coffee rust epidemic in Sri Lanka

1875 Mikhail Woronin identified the cause of clubroot as a "plasmodiophorous organism" and gave it the name *Plasmodiophora brassicae*

1876 *Fusarium oxysporum* f.sp. *cubense*, responsible for Panama disease, discovered in bananas in Australia

1878–1885 Downy mildew of grape epidemic in France

1879 Robert Koch establishes germ theory: diseases are caused by microorganisms

1882 *Lehrbuch der Baumkrankheiten* (*Textbook of Diseases of Trees*), by Robert Hartig, is published in Berlin, the first textbook of forest pathology.

1885 Bordeaux mixture introduced by Pierre-Marie-Alexis Millardet to control downy mildew on grape

1885 Experimental proof that bacteria can cause plant diseases: "Erwinia amylovora" and fire blight of apple

1886–1898 Recognition of plant viral diseases: Tobacco mosaic virus

1889 Introduction of hot water treatment of seed for disease control by Jensen

1902 First chair of plant pathology established, in Copenhagen

1904 Mendelian inheritance of cereal rust resistance demonstrated

1907 First academic department of plant pathology established, at Cornell University

1908 American Phytopathological Society founded

1910 Panama disease reaches Western Hemisphere

1911 Scientific journal *Phytopathology* founded

1925 Panama disease reaches every banana-growing country in the Western Hemisphere

1951 European and Mediterranean Plant Protection Organization (EPPO) founded

1967 Recognition of plant pathogenic mycoplasma-like organisms

1971 T. O. Diener discovers viroids, organisms smaller than viruses

The historical landmarks in plant pathology are taken from unless otherwise noted.

# Pathogen

In biology, a pathogen (Greek: *pathos* "suffering, passion" and "producer of") in the oldest and broadest sense is anything that can produce disease; the term came into use in the 1880s. Typically the term is used to describe an infectious agent such as a virus, bacterium, prion, a fungus, or even another micro-organism.

There are several substrates including *pathways* where the pathogens can invade a host. The principal pathways have different episodic time frames, but soil contamination has the longest or most persistent potential for harboring a pathogen. Diseases caused by organisms in humans are known as pathogenic diseases.

## Pathogenicity

Pathogenicity is the potential disease-causing capacity of pathogens. Pathogenicity is related to virulence in meaning, but some authorities have come to distinguish it as a *qualitative* term, whereas the latter is *quantitative*. By this standard, an organism may be said to be pathogenic or non-pathogenic in a particular context, but not "more pathogenic" than another. Such comparisons are described instead in terms of relative virulence. Pathogenicity is also distinct from the transmissibility of the virus, which quantifies the risk of infection.

A pathogen may be described in terms of its ability to produce toxins, enter tissue, colonize, hijack nutrients, and its ability to immunosuppress the host.

### Context-dependent Pathogenicity

It is common to speak of an entire species of bacteria as pathogenic when it is identified as the cause of a disease *(cf. Koch's postulates)*. However, the modern view is that pathogenicity depends on the microbial ecosystem as a whole. A bacterium may participate in opportunistic infections in immuno-compromised hosts, acquire virulence factors by plasmid infection, become transferred to a different site within the host, or respond to changes in the overall numbers of other bacteria present. For example, infection of mesenteric lymph glands of mice with *Yersinia* can clear the way for continuing infection of these sites by *Lactobacillus*, possibly by a mechanism of "immunological scarring".

### Related Concepts

### Virulence

Virulence (the tendency of a pathogen to cause damage to a host's fitness) evolves when that pathogen can spread from a diseased host, despite that host being very debilitated. Horizontal transmission occurs between hosts of the same species, in contrast to vertical transmission, which tends to evolve symbiosis (after a period of high morbidity and mortality in the population) by linking the pathogen's evolutionary success to the evolutionary success of the host organism.

Evolutionary medicine has found that under horizontal transmission, the host population might never develop tolerance to the pathogen.

### Transmission

Transmission of pathogens occurs through many different routes, including airborne, direct or indirect contact, sexual contact, through blood, breast milk, or other body fluids, and through the fecal-oral route.

### Types of Pathogens

### Bacterial

Although the vast majority of bacteria are harmless or beneficial, a relatively small list of pathogenic bacteria can cause infectious diseases. One of the bacterial diseases with the highest disease burden is tuberculosis, caused by the bacterium *Mycobacterium tuberculosis*, which kills about

2 million people a year, mostly in sub-Saharan Africa. Pathogenic bacteria contribute to other globally important diseases, such as pneumonia, which can be caused by bacteria such as *Streptococcus* and *Pseudomonas,* and foodborne illnesses, which can be caused by bacteria such as *Shigella, Campylobacter,* and *Salmonella.* Pathogenic bacteria also cause infections such as tetanus, typhoid fever, diphtheria, syphilis, and leprosy.

Bacteria can often be killed by antibiotics because the cell wall on the outside is destroyed, expelling the DNA out of the body of the pathogen, therefore making the pathogen incapable of producing proteins and dies. Bacteria typically range between 1 and 5 micrometers in length. A class of bacteria without cell walls is mycoplasma (a cause of lung infections). A class of bacteria which must live within other cells (obligate intracellular parasitic) is chlamydia (genus), the world leader in causing sexually transmitted infection (STD).

### Viral

Some of the diseases that are caused by viral pathogens include smallpox, influenza, mumps, measles, chickenpox, ebola, and rubella.

Pathogenic viruses are diseases mainly those of the families of: Adenoviridae, Picornaviridae, Herpesviridae, Hepadnaviridae, Flaviviridae, Retroviridae, Orthomyxoviridae, Paramyxoviridae, Papovaviridae, Polyomavirus, Rhabdoviridae, Togaviridae. Viruses typically range between 20-300 nanometers in length.

### Fungal

Fungi comprise a eukaryotic kingdom of microbes that are usually saprophytes (consume dead organisms) but can cause diseases in humans, animals and plants. Fungi are the most common cause of diseases in crops and other plants. The typical fungal spore size is 1-40 micrometers in length.

### Prionic

According to the prion theory, prions are infectious pathogens that do not contain nucleic acids. These abnormally folded proteins are found characteristically in some diseases such as scrapie, bovine spongiform encephalopathy (mad cow disease) and Creutzfeldt–Jakob disease.

### Other Parasites

Some eukaryotic organisms, such as protists and helminths, cause disease.

### Treatment and Health Care

Bacteria are usually treated with antibiotics while viruses are treated with antiviral compounds. Eukaryotic pathogens are typically not susceptible to antibiotics and thus need specific drugs. Infection with many pathogens can be prevented by immunization. A small amount of pathogens are used in vaccines to make immunity stay alert and strengthen defense on the insides to prepare for a larger quantity of the virus ever getting inside. Hygiene is critical for the prevention of infection by pathogens.

# References

- Jackson RW (editor). (2009). Plant Pathogenic Bacteria: Genomics and Molecular Biology. Caister Academic Press. ISBN 978-1-904455-37-0.

- Alberts B; Johnson A; Lewis J; et al. (2002). "Introduction to Pathogens". Molecular Biology of the Cell (4th ed.). Garland Science. p. 1. Retrieved 26 April 2016.

- "Plasmopara viticola, the Cause of Downy Mildew of Grapes". The Origin of Plant Pathology and The Potato Famine, and Other Stories of Plant Diseases. Retrieved 4 February 2015.

- "Fusarium oxysporum : The End of the Banana Industry?". The Origin of Plant Pathology and The Potato Famine, and Other Stories of Plant Diseases. Retrieved 4 February 2015.

- Nicole Davis (September 9, 2009). "Genome of Irish potato famine pathogen decoded". Haas et al. Broad Institute of MIT and Harvard. Retrieved 24 July 2012.

- "Scientists discover how deadly fungal microbes enter host cells". (VBI) at Virginia Tech affiliates. Physorg. July 22, 2010. Retrieved July 31, 2012.

- "1st large-scale map of a plant's protein network addresses evolution, disease process". Dana-Farber Cancer Institute. July 29, 2011. Retrieved 24 July 2012.

# 2

# Pathogens in Plant Pathology

Pathogens are what produce diseases. Generally, the term is used to describe an infectious agent such as virus, bacterium or a fungus. This chapter focuses on oomycete and the significance of oomycete plant pathogens and also focuses on bacteria and bacterial plant pathogens. The topics discussed in the chapter are of great importance to broaden the existing knowledge on plant pathology.

## Oomycete

Oomycota or oomycetes form a distinct phylogenetic lineage of fungus-like eukaryotic microorganisms. They are filamentous, microscopic, absorptive organisms that reproduce both sexually and asexually. Oomycetes occupy both saprophytic and pathogenic lifestyles, and include some of the most notorious pathogens of plants, causing devastating diseases such as late blight of potato and sudden oak death. One oomycete, the mycoparasite *Pythium oligandrum*, is used for biocontrol, attacking plant pathogenic fungi. The oomycetes are also often referred to as water molds (or water moulds), although the water-preferring nature which led to that name is not true of most species, which are terrestrial pathogens. The Oomycota have a very sparse fossil record. A possible oomycete has been described from Cretaceous amber.

### Morphology

The oomycetes rarely have septa, and if they do, they are scarce, appearing at the bases of sporangia, and sometimes in older parts of the filaments. Some are unicellular, but others are filamentous and branching.

### Phylogenetic Relationships

This group was originally classified among the fungi (the name "oomycota" means "egg fungus") and later treated as protists, based on general morphology and lifestyle. A cladistic analysis based on modern discoveries about the biology of these organisms supports a relatively close relationship with some photosynthetic organisms, such as brown algae and diatoms. A common taxonomic classification based on these data, places the class Oomycota along with other classes such as Phaeophyceae (brown algae) within the phylum Heterokonta.

This relationship is supported by a number of observed differences in the characteristics of oomycetes and fungi. For instance, the cell walls of oomycetes are composed of cellulose rather than chitin and generally do not have septations. Also, in the vegetative state they have diploid nuclei, whereas fungi have haploid nuclei. Most oomycetes produce self-motile zoospores with two flagella. One flagellum has a "whiplash" morphology, and the other a branched "tinsel" morphology. The "tinsel" flagellum is unique to the Kingdom Heterokonta. Spores of the few fungal groups which retain flagella (such as the Chytridiomycetes) have only one whiplash flagellum. Oomycota and fungi have different metabolic pathways for synthesizing lysine and have a number of enzymes that differ. The ultrastructure is also different, with oomycota having tubular mitochondrial cristae and fungi having flattened cristae.

In spite of this, many species of oomycetes are still described or listed as types of fungi and may sometimes be referred to as pseudofungi, or lower fungi.

## Classification

Previously the group was arranged into six orders.

- The Saprolegniales are the most widespread. Many break down decaying matter; others are parasites.

- The Leptomitales have wall thickenings that give their continuous cell body the appearance of septation. They bear chitin and often reproduce asexually.

- The Rhipidiales use rhizoids to attach their thallus to the bed of stagnant or polluted water bodies.

- The Albuginales are considered by some authors to be a family (Albuginaceae) within the Peronosporales, although it has been shown that they are phylogenetically distinct from this order.

- The Peronosporales too are mainly saprophytic or parasitic on plants, and have an aseptate, branching form. Many of the most damaging agricultural parasites belong to this order.

- The Lagenidiales are the most primitive; some are filamentous, others unicellular; they are generally parasitic.

However more recently this has been expanded considerably.

- Anisolpidiales Dick 2001
  - Anisolpidiaceae Karling 1943
- Lagenismatales Dick 2001
  - Lagenismataceae Dick 1995
- Salilagenidiales Dick 2001
  - Salilagenidiaceae Dick 1995
- Rozellopsidales Dick 2001
  - Rozellopsidaceae Dick 1995
  - Pseudosphaeritaceae Dick 1995
- Ectrogellales
  - Ectrogellaceae

- Haptoglossales
  - Haptoglossaceae
- Eurychasmales
  - Eurychasmataceae Petersen 1905
- Haliphthorales
  - Haliphthoraceae Vishniac 1958
- Olpidiopsidales
  - Sirolpidiaceae Cejp 1959
  - Pontismataceae Petersen 1909
  - Olpidiopsidaceae Cejp 1959
- Atkinsiellales
  - Atkinisellaceae
  - Crypticolaceae Dick 1995
- Saprolegniales
  - Achlyaceae
  - Verrucalvaceae Dick 1984
  - Saprolegniaceae Warm. 1884 [Leptolegniaceae]
- Leptomitales
  - Leptomitaceae Kuetz. 1843 [Apodachlyellaceae Dick 1986]
  - Leptolegniellaceae Dick 1971 [Ducellieriaceae Dick 1995]
- Rhipidiales
  - Rhipidiaceae Cejp 1959
- Albuginales
  - Albuginaceae Schroet. 1893
- Peronosporales [Pythiales; Sclerosporales; Lagenidiales]
  - Salisapiliaceae
  - Pythiaceae Schroet. 1893 [Pythiogetonaceae; Lagenaceae Dick 1994; Lagenidiaceae; Peronophythoraceae; Myzocytiopsidaceae Dick 1995]
  - Peronosporaceae Warm. 1884 [Sclerosporaceae Dick 1984]

## Phylogeny

## Etymology

"Oomycota" means "egg fungi", referring to the large round oogonia, structures containing the female gametes, that are characteristic of the oomycetes.

The name "water mold" refers to their earlier classification as fungi and their preference for conditions of high humidity and running surface water, which is characteristic for the basal taxa of the oomycetes.

## Biology

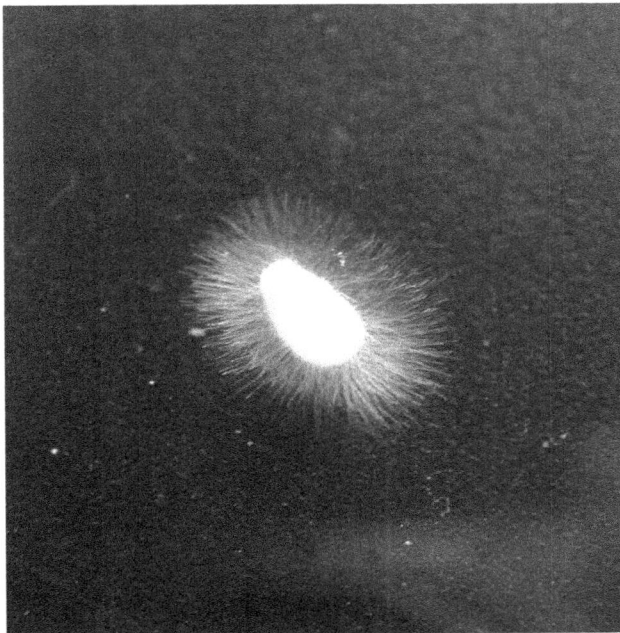

A culture of an oomycete from a stream

## Reproduction

Most of the oomycetes produce two distinct types of spores. The main dispersive spores are asexual, self-motile spores called zoospores, which are capable of chemotaxis (movement toward or away from a chemical signal, such as those released by potential food sources) in surface water (including precipitation on plant surfaces). A few oomycetes produce aerial asexual spores that are distributed by wind. They also produce sexual spores, called oospores, that are translucent, double-walled, spherical structures used to survive adverse environmental conditions.

## Pathogenicity

Many oomycetes species are economically important, aggressive plant pathogens. Some species can cause disease in fish. The majority of the plant pathogenic species can be classified into four groups, although more exist.

- The *Phytophthora* group is a paraphyletic genus that causes diseases such as dieback, late blight in potatoes (the cause of the Irish Potato Famine of the 1840s that ravaged Ireland and other parts of Europe), sudden oak death, rhododendron root rot, and ink disease in the European chestnut

- The paraphyletic *Pythium* group is more prevalent than *Phytophthora* and individual species have larger host ranges, although usually causing less damage. *Pythium* damping off is a very common problem in greenhouses, where the organism kills newly emerged seedlings. Mycoparasitic members of this group (e.g. *P. oligandrum*) parasitize other oomycetes and fungi, and have been employed as biocontrol agents. One *Pythium* species, *Pythium insidiosum*, is also known to infect mammals.

- The third group are the downy mildews, which are easily identifiable by the appearance of white, brownish or olive "mildew" on the leaf undersides (although this group can be confused with the unrelated fungal powdery mildews).

- The fourth group are the white blister rusts, Albuginales, which cause white blister disease on a variety of flowering plants. White blister rusts sporulate beneath the epidermis of their hosts, causing spore-filled blisters on stems, leaves and the inflorescence. The Albuginales are currently divided into three genera, *Albugo* parasitic predominantly to Brassicales, *Pustula*, parasitic predominantly to Asterales, and *Wilsoniana*, predominantly parasitic to Caryophyllales. Like the downy mildews, the white blister rusts are obligate biotrophs, which means that they are unable to survive without the presence of a living host.

## Significant Oomycete Plant Pathogens

## Pythium

*Pythium* is a genus of parasitic oomycotes. Most species are plant parasites, but *Pythium insidiosum* is an important pathogen of animals. They were formerly classified as fungi; the feet of the fungus gnat are frequently a vector for their transmission.

## Morphology

Hyphae

> *Pythium* species, like others in the family Pythiaceae, are usually characterized by their production of coenocytic hyphae without septations.

Oogonia

> Generally contain a single oospore

Antheridia

> Contain an elongated and club-shaped antheridium

## Ecological Importance

*Pythium*-induced root rot is a common crop disease. When the organism kills newly emerged or emerging seedlings, it is known as damping off, and is a very common problem in fields and greenhouses. This disease complex usually involves other pathogens such as *Phytophthora* and *Rhizoctonia*. Pythium wilt is caused by zoospore infection of older plants, leading to biotrophic infections that become necrotrophic in response to colonization/reinfection pressures or environmental stress, leading to minor or severe wilting caused by impeded root functioning.

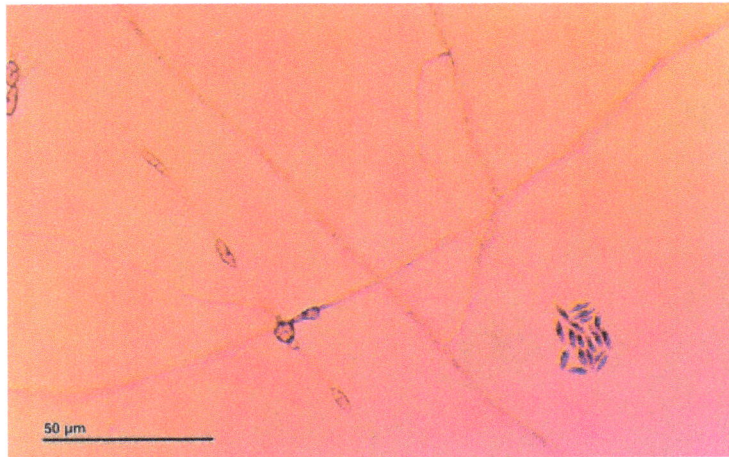

Pythium

Many *Pythium* species, along with their close relatives *Phytophthora*, are plant pathogens of economic importance in agriculture. *Pythium* spp. tend to be very generalistic and unspecific in their large range of hosts, while *Phytophthora* spp. are generally more host-specific.

For this reason, *Pythium* spp. are more devastating in the root rot they cause in crops, because crop rotation alone often does not eradicate the pathogen (nor does fallowing the field, as *Pythium* spp. are also good saprotrophs, and survive for a long time on decaying plant matter).

In field crops, damage by *Pythium* spp. is often limited to the area affected, as the motile zoospores require ample surface water to travel long distances. Additionally, the capillaries formed by soil particles act as a natural filter and effectively trap many zoospores. However, in hydroponic systems inside greenhouses, where extensive monocultures of plants are maintained in plant nutrient solution (containing nitrogen, potassium, phosphate, and micronutrients) that is continuously recirculated to the crop, *Pythium* spp. cause extensive and devastating root rot and is often difficult to prevent or control. The root rot affects entire operations (tens of thousands of plants, in many instances) within two to four days due to the inherent nature of hydroponic systems where roots are nakedly exposed to the water medium, in which the zoospores can move freely.

Several *Pythium* species, including *P. oligandrum*, *P. nunn*, *P. periplocum*, and *P. acanthicum*, are mycoparasites of plant pathogenic fungi and oomycetes, and have received interest as potential biocontrol agents.

## Phytophthora

*Phytophthora* (from Greek (*phytón*), "plant" and (*phthorá*), "destruction"; "the plant-destroyer") is a genus of plant-damaging Oomycetes (water molds), whose member spe-cies are capable of causing enormous economic losses on crops worldwide, as well as environ-mental damage in natural ecosystems. The cell wall of *Phytophthora* is made up of cellulose. The genus was first described by Heinrich Anton de Bary in 1875. Approximately 100 species have been described, although 100–500 undiscovered *Phytophthora* species are estimated to exist.

Sudden oak death caused by *Phytophthora ramorum*

## Pathogenicity

*Phytophthora* spp. are mostly pathogens of dicotyledons, and are relatively host-specific para-sites. Many species of *Phytophthora* are plant pathogens of considerable economic importance. *Phytophthora infestans* was the infective agent of the potato blight that caused the Great Irish Famine (1845–1849), and still remains the most destructive pathogen of solanaceous crops, in-cluding tomato and potato. The soya bean root and stem rot agent, *Phytophthora sojae*, has also caused longstanding problems for the agricultural industry. In general, plant diseases caused by this genus are difficult to control chemically, thus the growth of resistant cultivars is the main management strategy. Other important *Phytophthora* diseases are:

- *Phytophthora* taxon Agathis—causes collar-rot on New Zealand kauri (*Agathis australis*), New Zealand's most voluminous tree, an otherwise successful survivor of the Jurassic

- *Phytophthora cactorum*—causes rhododendron root rot affecting rhododendrons, azaleas and causes bleeding canker in hardwood trees

- *Phytophthora capsici*—infects Cucurbitaceae fruits, such as cucumbers and squash

- *Phytophthora cinnamomi*—causes cinnamon root rot affecting woody ornamentals includ-ing arborvitae, azalea, Chamaecyparis, dogwood, forsythia, Fraser fir, hemlock, Japanese holly, juniper, Pieris, rhododendron, Taxus, white pine, American chestnut and Australian Jarrah.

- *Phytophthora fragariae*—causes red root rot affecting strawberries

- *Phytophthora kernoviae*—pathogen of beech and rhododendron, also occurring on other trees and shrubs including oak, and holm oak. First seen in Cornwall, UK, in 2003.

- *Phytophthora megakarya*—one of the cocoa black pod disease species, is invasive and probably responsible for the greatest cocoa crop loss in Africa

- *Phytophthora nicotianae*—infects onions

- *Phytophthora palmivora*—causes fruit rot in coconuts and betel nuts

- *Phytophthora ramorum*—infects over 60 plant genera and over 100 host species; causes sudden oak death

- *Phytophthora quercina*—causes oak death

- *Phytophthora sojae*—causes soybean root rot

Research beginning in the 1990s has placed some of the responsibility for European forest die-back on the activity of imported Asian *Phytophthoras*.

## Fungi Resemblance

*Phytophthora* is sometimes referred to as a fungus-like organism, but it is classified under a different kingdom altogether: Chromalveolata (formerly Stramenopila and previously Chromista). This is a good example of convergent evolution: *Phytophthora* is morphologically very similar to true fungi yet its evolutionary history is quite distinct. In contrast to fungi, chromalveolatas are more closely related to plants than animals. Whereas fungal cell walls are made primarily of chitin, chromalveolata cell walls are constructed mostly of cellulose. Ploidy levels are different between these two groups; *Phytophthora* species have diploid (paired) chromosomes in the vegetative (growing, nonreproductive) stage of life, whereas fungi are almost always haploid in this stage. Biochemical pathways also differ, notably the highly conserved lysine synthesis path.

## Biology

*Phytophthora* species may reproduce sexually or asexually. In many species, sexual structures have never been observed, or have only been observed in laboratory matings. In homothallic species, sexual structures occur in single culture. Heterothallic species have mating strains, designated as A1 and A2. When mated, antheridia introduce gametes into oogonia, either by the oogonium passing through the antheridium (amphigyny) or by the antheridium attaching to the proximal (lower) half of the oogonium (paragyny), and the union producing oospores. Like animals, but not like most true fungi, meiosis is gametic, and somatic nuclei are diploid. Asexual (mitotic) spore types are chlamydospores, and sporangia which produce zoospores. Chlamydospores are usually spherical and pigmented, and may have a thickened cell wall to aid in their role as a survival structure. Sporangia may be retained by the subtending hyphae (noncaducous) or be shed readily by wind or water tension (caducous) acting as dispersal structures. Also, sporangia may release zoospores, which have two unlike flagella which they use to swim towards a host plant.

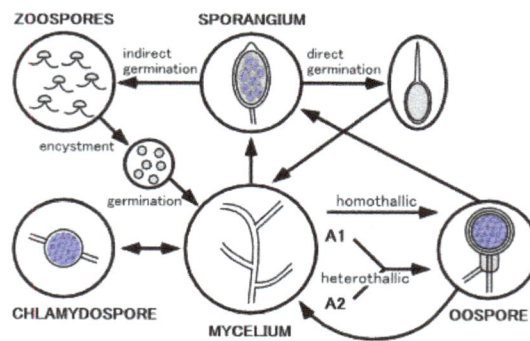

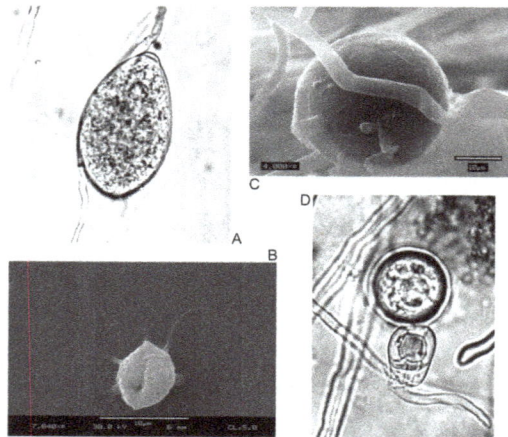

*Phytophthora* forms: A: Sporangia. B: Zoospore. C: Chlamydospore. D: Oospore

## Phytophthora Infestans

*Phytophthora infestans* is an oomycete that causes the serious potato disease known as late blight or potato blight. (Early blight, caused by *Alternaria solani*, is also often called "potato blight".) Late blight was a major culprit in the 1840s European, the 1845 Irish and 1846 Highland potato famines. The organism can also infect tomatoes and some other members of the Solanaceae. At first, the spots are gray-green and water-soaked, but they soon enlarge and turn dark brown and firm, with a rough surface.

## Etymology

The genus name *Phytophthora* comes from the Greek, meaning "plant", plus the Greek, meaning "destruction, ruin". The species name *infestans* is the present participle of the Latin verb *infestare*, meaning "attacking, destroying".

## Biology

The asexual life cycle of *Phytophthora infestans* is characterized by alternating phases of hyphal growth, sporulation, sporangia germination (either through zoospore release or direct germination, i.e. germ tube emergence from the sporangium), and the re-establishment of hyphal growth. There is also a sexual cycle, which occurs when isolates of opposite mating type (A1 and A2) meet. Hormonal communication triggers the formation of the sexual spores, called oospores. The different types of spores play major roles in the dissemination and survival of *P. infestans*. Sporangia are spread by wind or water and enable the movement of *P. infestans* between different host plants. The zoospores released from sporangia are biflagellated and chemotactic, allowing further movement of *P. infestans* on water films found on leaves or soils. Both sporangia and zoospores are short-lived, in contrast to oospores which can persist in a viable form for many years.

Under ideal conditions, the life cycle can be completed on potato or tomato foliage in about five days. Sporangia develop on the leaves, spreading through the crop when temperatures are above 10 °C (50 °F) and humidity is over 75%-80% for 2 days or more. Rain can wash spores into the soil where they infect young tubers, and the spores can also travel long distances on the wind. The early stages of blight are easily missed. Symptoms include the appearance of dark blotches on leaf tips and plant stems. White mould will appear under the leaves in humid conditions and the whole plant may quickly collapse. Infected tubers develop grey or dark patches that are reddish brown beneath the skin, and quickly decay to a foul-smelling mush caused by the infestation of secondary soft bacterial rots. Seemingly healthy tubers may rot later when in store.

*P. infestans* survives poorly in nature apart from its plant hosts. Under most conditions, the hyphae and asexual sporangia can survive for only brief periods in plant debris or soil, and are generally killed off during frosts or very warm weather. The exceptions involve oospores, and hyphae present within tubers. The persistence of viable pathogen within tubers, such as those that are left in the ground after the previous year's harvest or left in cull piles is a major problem in disease management. In particular, volunteer plants sprouting from infected tubers are thought to be a major source of inoculum at the start of a growing season. This can have devastating effects by destroying entire crops.

## Genetics

*P. infestans* is diploid, with about 11-13 chromosomes, and in 2009 scientists completed the sequencing of its genome. The genome was found to be considerably larger (240 Mbp) than that of most other *Phytophthora* species whose genomes have been sequenced; *Phytophthora sojae* has a 95 Mbp genome and *Phytophthora ramorum* had a 65 Mbp genome. About 18,000 genes were detected within the *P. infestans* genome. It also contained a diverse variety of transposons and many gene families encoding for effector proteins that are involved in causing pathogenicity. These proteins are split into two main groups depending on whether they are produced by the water mould in the symplast (inside plant cells) or in the apoplast (between plant cells). Proteins produced in the symplast included RXLR proteins, which contain an arginine-X-leucine-arginine (where X can be any amino acid) sequence at the amino terminus of the protein. Some RXLR proteins are avirulence proteins, meaning that they can be detected by the plant and lead to a hypersensitive response which restricts the growth of the pathogen. *P. infestans* was found to encode around 60% more of these proteins than most other *Phytophthora* species. Those found in the apoplast include hydrolytic enzymes such as proteases, lipases and glycosylases that act to degrade plant tissue, enzyme inhibitors to protect against host defence enzymes and necrotizing toxins. Overall the genome was found to have an extremely high repeat content (around 74%) and to have an unusual gene distribution in that some areas contain many genes whereas others contain very few.

## Origin and Diversity of P. Infestans

Potatoes infected with late blight are shrunken on the outside, corky and rotted inside.

Historical model of *Phytophthora infestans*, Botanical Museum Greifswald

Historical model of a potato leaf with *Phytophthora infestans*, Botanical Museum Greifswald

The highlands of central Mexico are considered by many to be the center of origin of *P. infestans*, although others have proposed its origin to be in the Andes, which is also the origin of potatoes. A recent study evaluated these two alternate hypotheses and found conclusive support for central Mexico being the center of origin. Support for Mexico comes from multiple observations including the fact that populations are genetically most diverse in Mexico, late blight is observed in native tuber-bearing *Solanum* species, populations of the pathogen are in Hardy-Weinberg equilibrium, the two mating types occur in a 1:1 ratio, and detailed phylogeographic and evolutionary studies. Furthermore, the closest relatives of *P. infestans*, namely *P. mirabilis* and *P. ipomoeae* are endemic to central Mexico. On the other hand, the only close relative found in South America, namely *P. andina*, is a hybrid that does not share a single common ancestor with *P. infestans*. Finally, populations of *P. infestans* in South America lack genetic diversity and are clonal.

The Great Irish Famine was largely due to the biggest hits with Phytophthora Infestans. It lasted from 1845 to 1851 and it killed more than 174,000 people.

Migrations from Mexico to North America or Europe have occurred several times throughout history, probably linked to the movement of tubers. Until the 1970s, the A2 mating type was restricted to Mexico, but now in many regions of the world both A1 and A2 isolates can be found in the same region. The co-occurrence of the two mating types is significant due to the possibility of sexual recombination and formation of oospores, which can survive the winter. Only in Mexico and Scandinavia, however, is oospore formation thought to play a role in overwintering. In other parts of Europe, increasing genetic diversity has been observed as a consequence of sexual reproduction This is notable since different forms of *P. infestans* vary in their aggressiveness on potato or tomato, in sporulation rate, and sensitivity to fungicides. Variation in such traits also occurs in North America, however importation of new genotypes from Mexico appears to be the predominant cause of genetic diversity, as opposed to sexual recombination within potato or tomato fields. Many of the strains that appeared outside of Mexico since the 1980s have been more aggressive, leading to increased crop losses. Some of the differences between strains may be related to variation in the RXLR effectors that are present.

## Disease Management

*P. infestans* is still a difficult disease to control. There are many chemical options in agriculture for the control of both damage to the foliage and infections of the tuber. A few of the most common foliar-applied fungicides are Ridomyl, a Gavel/SuperTin tank mix, and Previcur Flex. Orondis is a new product from Syngenta that shows promise but availability in 2016 will be limited. All of the aforementioned fungicides need to be tank mixed with a broad-spectrum fungicide such as mancozeb or chlorothalonil not just for resistance management but also because the potato plants will be attacked by other pathogens at the same time.

If adequate field scouting occurs and late blight is found soon after disease development, localized patches of potato plants can be killed with a dessicant (e.g. paraquat) through the use of a backpack sprayer. This management technique can be thought of as a field-scale hypersensitive response similar to what occurs in some plant-viral interactions whereby cells surrounding the initial point of infection are killed in order to prevent proliferation of the pathogen.

If infected tubers make it into the storage bin, there's a very high risk to the storage life of that bin. Once in storage, there isn't much that can be done besides emptying the parts of the bin that contain tubers infected with *Phytophthora infestans*. To increase the probability of successfully storing potatoes from a field where late blight was known to occur during the growing season, some products can be applied just prior to entering storage (e.g. Phostrol). The problem with products being sprayed on tubers just prior to storage is that you are applying these products in an aqueous solution and high moisture carries a high risk of tuber breakdown due to wide range of pathogens.

Around the world the disease causes around $6 billion of damage to crops each year.

## Resistant Plants

Breeding for resistance, particularly in potato, has had limited success in part due to difficulties in crossing cultivated potato to its wild relatives, which are the source of potential resistance genes. In addition, most resistance genes only work against a subset of *P. infestans* isolates, since effective plant disease resistance only results when the pathogen expresses a RXLR effector gene that matches the corresponding plant resistance (R) gene; effector-R gene interactions trigger a range of plant defenses, such as the production of compounds toxic to the pathogen.

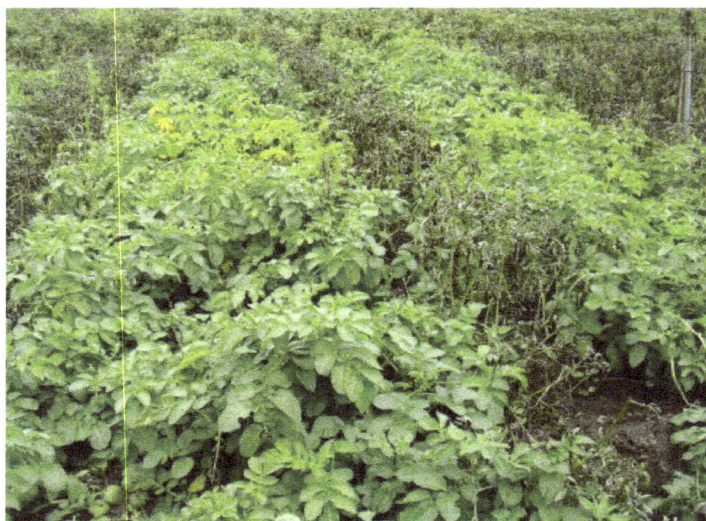

Potatoes after exposure to *Phytophthora infestans*. The normal potatoes have blight but the cisgenic potatoes are healthy.

Potato and tomato varieties vary in their susceptibility to blight. Most early varieties are very vulnerable; they should be planted early so that the crop matures before blight starts (usually in July in the Northern Hemisphere). Many old crop varieties, such as King Edward potato are also very susceptible but are grown because they are wanted commercially. Maincrop varieties which are very slow to develop blight include Cara, Stirling, Teena, Torridon, Remarka, and Romano. Some so-called resistant varieties can resist some strains of blight and not others, so their performance may vary depending on which are around. These crops have had polygenic resistance bred into them, and are known as "field resistant". New varieties such as Sarpo Mira and Sarpo Axona show great resistance to blight even in areas of heavy infestation. Defender is an American cultivar whose parentage includes Ranger Russet and Polish potatoes resistant to late blight. It is a long white-skinned cultivar with both foliar and tuber resistance to late blight. Defender was released in 2004.

Genetic engineering may also provide options for generating resistance cultivars. A resistance gene effective against most known strains of blight has been identified from a wild relative of the potato, *Solanum bulbocastanum*, and introduced by genetic engineering into cultivated varieties of potato. This is an example of cisgenic genetic engineering.

## Reducing Inoculum

Blight can be controlled by limiting the source of inoculum. Only good-quality seed potatoes and tomatoes obtained from certified suppliers should be planted. Often discarded potatoes from the previous season and self-sown tubers can act as sources of inoculum.

## Environmental Conditions

There are several environmental conditions that are conducive to *P. infestans*. An example of such took place in the United States during the 2009 growing season. As colder than average for the season and with greater than average rainfall, there was a major infestation of tomato plants, specifically in the eastern states. By using weather forecasting systems, such as BLITECAST, if the following conditions occur as the canopy of the crop closes, then the use of fungicides is recommended to prevent an epidemic.

- A Beaumont Period is a period of 48 consecutive hours, in at least 46 of which the hourly readings of temperature and relative humidity at a given place have not been less than 10 °C (50 °F) and 75%, respectively.

- A Smith Period is at least two consecutive days where min temperature is 10 °C (50 °F) or above and on each day at least 11 hours when the relative humidity is greater than 90%.

The Beaumont and Smith periods have traditionally been used by growers in the United Kingdom, with different criteria developed by growers in other regions. The Smith period has been the preferred system used in the UK since its introduction in the 1970s.

Based on these conditions and other factors, several tools have been developed to help growers manage the disease and plan fungicide applications. Often these are deployed as part of Decision Support Systems accessible through web sites or smart phones.

## Use of Fungicides

Spraying in a potato field for prevention of potato blight in Nottinghamshire, England.

Fungicides for the control of potato blight are normally only used in a preventative manner, optionally in conjunction with disease forecasting. In susceptible varieties, sometimes fungicide applications may be needed weekly. An early spray is most effective. The choice of fungicide can depend on the nature of local strains of *P. infestans*. Metalaxyl is a fungicide that was marketed for use against *P. infestans*, but suffered serious resistance issues when used on its own. In some regions of the world during the 1980s and 1990s, most strains of *P. infestans* became resistant to metalaxyl, but in subsequent years many populations shifted back to sensitivity. To reduce the occurrence of resistance, it is strongly advised to use single-target fungicides such as metalaxyl along with carbamate compounds. A combination of other compounds are recommended for managing metalaxyl-resistant strains. These include mandipropamid, chlorothalonil, fluazinam, triphenytin, mancozeb and others. In the past, copper sulfate solution (called 'bluestone') was used to combat potato blight. Copper pesticides remain in use on organic crops, both in the form of copper hydroxide and copper sulfate. Given the dangers of copper toxicity, other organic control options that have been shown to be effective include horticultural oils, phosphorous acids, and rhamnolipid biosurfactants, while sprays containing "beneficial" microbes such as *Bacillus subtilis* or compounds that encourage the plant to produce defensive chemicals (such as knotweed extract) have not performed as well.

## Control of Tuber Blight

Ridging is often used to reduce tuber contamination by blight. This normally involves piling soil or mulch around the stems of the potato blight meaning the pathogen has farther to travel to get to the tuber. Another approach is to destroy the canopy around five weeks before harvest, using a contact herbicide or sulfuric acid to burn off the foliage. By eliminating infected foliage, this reduces the likelihood of tuber infection.

## Historical Impact

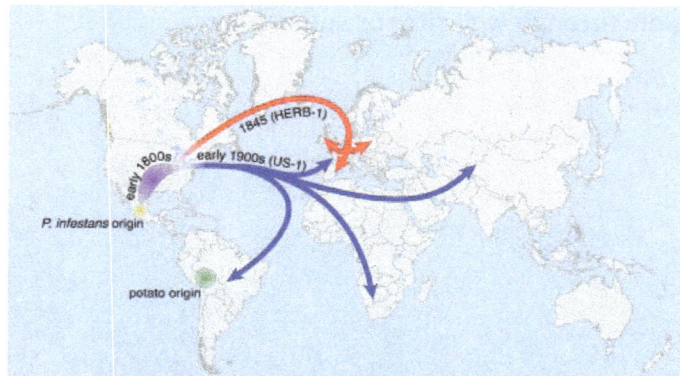

Suggested paths of migration and diversification of P. infestans lineages HERB-1 and US-1

The effects of *Phytophthora infestans* in Ireland in 1845–57 were one of the factors which caused over one million to starve to death and forced another two million to emigrate from affected countries. Most commonly referenced is the Great Irish Famine, during the late 1840s. The first recorded instances of the disease were in the United States, in Philadelphia and New York City in early 1843. Winds then spread the spores, and in 1845 it was found from Illinois to Nova Scotia, and from Virginia to Ontario. It crossed the Atlantic Ocean with a shipment of seed potatoes for Belgian farmers in 1845. All of the potato-growing countries in Europe were affected, but the potato blight hit Ireland the hardest. Implicated in Ireland's fate was the island's disproportionate dependency on a single variety of potato, the Irish Lumper. The lack of genetic variability created a susceptible host population for the organism.

During the First World War, all of the copper in Germany was used for shell casings and electric wire and therefore none was available for making copper sulfate to spray potatoes. A major late blight outbreak on potato in Germany therefore went untreated, and the resulting scarcity of potatoes led to the deaths of 700,000 German civilians from starvation.

France, Canada, the United States, and the Soviet Union researched *P. infestans* as a biological weapon in the 1940s and 1950s. Potato blight was one of more than 17 agents that the United

States researched as potential biological weapons before the nation suspended its biological weapons program. Whether a weapon based on the pathogen would be effective is questionable, due to the difficulties in delivering viable pathogen to an enemy's fields, and the role of uncontrollable environmental factors in spreading the disease.

## Phytophthora Ramorum

*Phytophthora ramorum* is the oomycete plant pathogen known to cause the disease sudden oak death (SOD). The disease kills oak and other species of trees and has had devastating effects on the oak populations in California and Oregon, as well as being present in Europe. Symptoms include bleeding cankers on the tree's trunk and dieback of the foliage, in many cases eventually leading to the death of the tree.

*P. ramorum* also infects a great number of other plant species, significantly woody ornamentals such as *Rhododendron, Viburnum*, and *Pieris*, causing foliar symptoms known as ramorum dieback or ramorum blight. Such plants can act as a source of inoculum for new infections, with the pathogen-producing spores that can be transmitted by rainsplash and rainwater.

*P. ramorum* was first reported in 1995, and the origins of the pathogen are still unclear, but most evidence suggests it was repeatedly introduced as an exotic species. Very few control mechanisms exist for the disease, and they rely upon early detection and proper disposal of infected plant material.

## Presence

The disease is known to exist in California's coastal region between Big Sur (in Monterey County) and southern Humboldt County. It is confirmed to exist in all coastal counties in this range, as well as in all immediately inland counties from Santa Clara County north to Lake County. It has not been found east of the California Coast Ranges, however. It was reported in Curry County, Oregon (just north of the California border), in 2001. Sonoma County has been hit hardest, having more than twice the area of new mortality of any other county in California.

About the same time, a similar disease in continental Europe and the UK was also identified as *Phytophthora ramorum*.

## Hosts and Symptoms

### In North America

It was first discovered in California in 1995 when large numbers of tanoaks (*Lithocarpus densiflorus*) died mysteriously, and was described as a new species of *Phytophthora* in 2000. It has subsequently been found in many other areas, including Britain, Germany, and some other U.S. states, either accidentally introduced in nursery stock, or already present undetected.

In tanoaks, the disease may be recognized by wilting new shoots, older leaves becoming pale green, and after a period of two to three weeks, foliage turning brown while clinging to the branches. Dark brown sap may stain the lower trunk's bark. Bark may split and exude gum, with visible discoloration. After the tree dies back, suckers try to sprout the next year, but their tips soon bend and die.

Ambrosia beetles (*Monarthrum scutellare*) will most likely infest a dying tree during midsummer, producing piles of fine white dust near tiny holes. Later, bark beetles (*Pseudopityophthorus pubipennis*) produce fine, red boring dust. Small black domes, the fruiting bodies of the *Hypoxylon* fungus, may also be present on the bark. Leaf death may occur more than a year after the initial infection and months after the tree has been girdled by beetles.

A hillside in Big Sur, California, devastated by sudden oak death

In coast live oaks and Californian black oaks, the first symptom is a burgundy-red to tar-black thick sap bleeding from the bark surface. These are often referred to as bleeding cankers.

In addition to oaks, many other forest species may be hosts for the disease; in fact, it was observed in the USA that nearly all woody plants in some Californian forests were susceptible to *P. ramorum*. including rhododendron, madrone (*Arbutus menziesii*), evergreen huckleberry (*Vaccinium ovatum*), California bay laurel (*Umbellularia californica*), buckeye (*Aesculus californica*), bigleaf maple (*Acer macrophyllum*), toyon (*Heteromeles arbutifolia*), manzanita (*Arctostaphylos spp.*), coast redwood (*Sequoia sempervirens*), Douglas fir (*Pseudotsuga menziesii*), coffeeberry (*Rhamnus californica*), honeysuckle (*Lonicera hispidula*), and Shreve oak (*Quercus parvula*). *P. ramorum* more commonly causes a less severe disease known as ramorum dieback/leaf blight on these hosts. Characteristic symptoms are dark spots on foliage and in some hosts the dieback of the stems and twigs. The disease is capable of killing some hosts, such as rhododendron, but most survive. Disease progression on these species is not well documented. Redwoods exhibit needle discoloration and cankers on small branches, with purple lesions on sprouts that may lead to sprout mortality.

## In Europe

In Europe, *Ramorum* blight was first observed on rhododendron and viburnum in the early 1990s, where it was initially found mainly on container-grown plants in nurseries. The principal symptoms were leaf and twig blight. By 2007, it had spread throughout nurseries and retail centers in 16 European countries, and had been detected in gardens, parks, and woodlands in at least eight countries. It has not caused significant harm to European oak species.

In 2009, the pathogen was found to be infecting and killing large numbers of Japanese larch trees (*Larix kaempferi*) in the United Kingdom at sites in the English counties of Somerset, Devon, and Cornwall. It was the first time in the world that *Phytophthora ramorum* had been found infecting this species. Since then, it has also been found extensively in larch plantations in Wales and in southwest Scotland, leading to the deliberate felling of large areas. The UK Forestry Commission noted that eradication of the disease would not be possible, and instead adopted a strategy of containing the disease to reduce its spread. Symptoms of the disease on larch trees include dieback of the tree's crown and branches, and a distinctive yellowing or ginger colour beneath the bark. In August 2010, disease was found in Japanese larch trees, in Counties Waterford and Tipperary in Ireland.

Leaf death caused by *P. ramorum*

The closely related *Phytophthora kernoviae* causes similar symptoms to *P. ramorum*, but infects the European beech (*Fagus sylvatica*).

## Transmission

*P. ramorum* produces both resting spores (chlamydospores) and zoospores, which have flagella enabling swimming. *P. ramorum* is spread by air; one of the major mechanisms of dispersal is rainwater splashing spores onto other susceptible plants, and into watercourses to be carried for greater distances. Chlamydospores can withstand harsh conditions and are able to overwinter. The pathogen will take advantage of wounding, but it is not necessary for infection to occur.

As mentioned above, *P. ramorum* does not kill every plant that can be used as a host, and these plants are most important in the epidemiology of the disease as they act as sources of inoculum. In the USA, bay laurel seems to be the main source of inoculum in forests. Green waste, such as leaf litter and tree stumps, are also capable of supporting *P. ramorum* as a saprotroph and acting as a source of inoculum. Because *P. ramorum* is able to infect many ornamental plants, it can be transmitted by ornamental plant movement.

*Cannabis* cultivation and associated traffic and movement of supplies and soil amendments in Northern California watersheds correlate with areas of introduction of *P. ramorum*. Hikers, mountain bikers, equestrians, and other people engaged in various outdoor activities may also unwittingly move the pathogen into areas where it was not previously present. Those travelling in an area known to be infested with SOD can help prevent the spread of the disease by cleaning their (and their animals') feet, tires, tools, camping equipment, etc. before returning home or entering another uninfected area, especially if they have been in muddy soil. Additionally, the movement of firewood could introduce sudden oak death to otherwise uninfected areas. Both homeowners and travelers are advised to buy and burn local firewood.

## The Two Mating Types

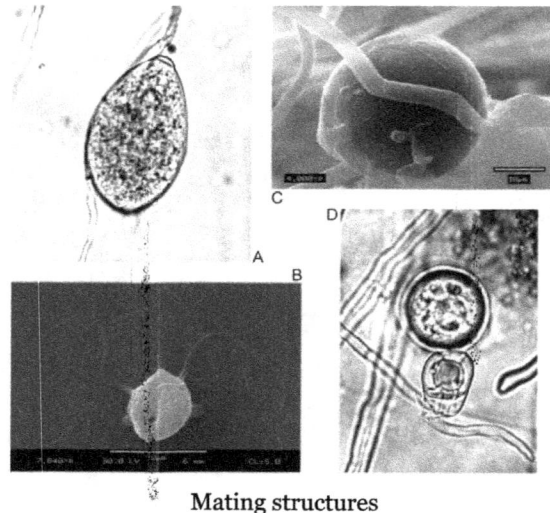

Mating structures

*P. ramorum* is heterothallic and has two mating types, A1 and A2, required for sexual reproduction. Interestingly, the European population is predominantly A1 while both mating types A1 and A2 are found in North America. Genetics of the two isolates indicate that they are reproductively isolated. On average, the A1 mating type is more virulent than the A2 mating type, but more variation occurs in the pathogenicity of A2 isolates. It is currently not clear whether this pathogen can reproduce sexually in nature and genetic work has suggested that the lineages of the two mating types might be isolated reproductively or geographically given the evolutionary divergence observed.

## Possible Origins

*P. ramorum* is a relatively new disease, and several debates have occurred about where it may have originated or how it evolved.

## Introduction As An Exotic Species

Evidence suggests *P. ramorum* may be an introduced species, and these introductions occurred separately for the European and North American populations, hence why only one mating type exists on each continent – this is called a founder effect. The differences between the two populations are thus caused by adaptation to separate climates. Evidence includes little genetic variability, as *P. ramorum* has not had time to diversify since being introduced. Existing variability may be explained by multiple introductions with a few individuals adapting best to their respective environments. The behavior of the pathogen in California is also indicative of being introduced; it is assumed that such a high mortality rate of trees would have been noticed sooner if *P. ramorum* were native.

Where *P. ramorum* originated remains unclear, but most researchers feel Asia is the most likely, since many of the hosts of *P. ramorum* originated there. Since certain climates are best suited to *P. ramorum*, the most likely sources are the southern Himalayas, Tibet, or Yunnan province.

## Hybridization Events

Species of *Phytophthora* have been shown to have evolved by the interspecific hybridization of two different species from the genus. When a species is introduced into a new environment, it causes episodic selection. The invading species is exposed to other resident taxa, and hybridization may occur to produce a new species. If these hybrids are successful, they may outcompete their parent species. Thus, *P, ramorum* is possibly a hybrid between two species. Its unique morphology does support this. Also, three sequences studied to establish the phylogeny of *Phytophthora*: ITS, cox II and nad 5, were identical, supporting *P. ramorum* having recently evolved.

## A native Organism

*P. ramorum* may be native to the United States. Infection rates could have previously been at a low level, but changes in the environment caused a change to the population structure. Alternatively, the symptoms of *P. ramorum* may have been mistaken for that of other pathogens. When SOD first appeared in the United States, many other pathogens and conditions were blamed before *P. ramorum* was found to be the causal agent. With many of the most seriously affected plants being in the forest, the likelihood of seeing diseased trees is also low.

## Ecological Impacts

In relation to human ecology, the loss of tanoak as the pathogen spreads to culturally sensitive indigenous lands represents a loss of tanoak acorn as one of the most important traditional and ceremonial foods still used in Northern California such as among Yurok people, Hupa, Miwok, and Karuk peoples. Similar impact applies to the decline of other native plant species that are traditional food sources in tanoak and oak regimens infected by the pathogen.

In forest ecology, the pathogen contributes to loss of environmental services provided by diversity of plant species and interdependent wildlife.

The mortality caused by this emerging disease is expected to cause many indirect effects. Several predictions of long-term impacts have been discussed in the scientific literature. While such predictions are necessarily speculative, indirect impacts occurring on shorter time scales have been documented in a few cases. For instance, one study demonstrated that redwood trees (*Sequoia sempervirens*) grew faster after neighboring tanoaks were killed by sudden oak death. Other studies have combined current observations and reconstruction/projection techniques to document short-term impacts while also inferring future conditions. One study used this approach to investigate the effects of SOD on the structural characteristics of redwood forests.

Additional long-term impacts of SOD may be inferred from regeneration patterns in areas that have experienced severe mortality. These patterns may indicate which tree species will replace tanoak in diseased areas. Such transitions will be of particular importance in forest types that were relatively poor in tree species diversity before the introduction of SOD, e.g., redwood forest. As of 2011, the only study to comprehensively examine regeneration in SOD-impacted redwood forests found no evidence that other broadleaf tree species are beginning to recruit. Instead, redwood was colonizing most mortality gaps. However, they also found inadequate regeneration in some areas and concluded that regeneration is continuing. Since this study only considered one site in Marin County, California, these results may not apply to other forests. Other impacts to the local ecology include, among others, the residual effects of spraying heavy pesticides (Agrifos) to treat SOD symptoms, and the heavy mortality of the native pollinator community that occurs as a result. Bee hives situated in areas of heavy Agrifos spraying have incurred significant losses of population in direct correlation to the application of these chemicals. Counties such as Napa and Sonoma may be doing significant damage to their native pollinator populations by virtue of adopting broad-based prophylactic pesticide policies. Such damage to the pollinator populations may have tertiary negative effects on the entire local plant community, compounding the loss of biodiversity, and thus environmental value, attributable to SOD.

## Control

### Early Detection

Early detection of *P. ramorum* is essential for its control. On an individual-tree basis, preventive treatments, which are more effective than therapeutic treatments, depend on knowledge of the pathogen's movement through the landscape to know when it is nearing prized trees. On the landscape level, *P. ramorum's* fast and often undetectable movement means that any treatment hoping to slow its spread must happen very early in the development of an infestation. Since *P. ramorum's* discovery, researchers have been working on the development of early detection methods on scales ranging from diagnosis in individual infected plants to landscape-level detection efforts involving large numbers of people.

Detecting the presence of *Phytophthora* species requires laboratory confirmation. The traditional method of culturing is on a growth medium selective against fungi (and, in some cases, against other oomycetes such as *Pythium* species). Host material is removed from the leading edge of a plant tissue canker caused by the pathogen; resulting growth is examined under a microscope to confirm the unique morphology of *P. ramorum*. Successful isolation of the pathogen often depends on the type of host tissue and the time of year that detection is attempted.

Because of these difficulties, researchers have developed some other approaches for identifying *P. ramorum*. The enzyme-linked immunosorbent assay test can be the first step in nonculture methods of identifying *P. ramorum*, but it can only be a first step, because it detects the presence of proteins that are produced by all *Phytophthora* species. In other words, it can identify to the genus level, but not to the species level. ELISA tests can process large numbers of samples at once, so researchers often use it to screen out likely positive samples from those that are not when the total number of samples is very large. Some manufacturers produce small-scale ELISA "field kits" that the homeowner can use to determine if plant tissue is infected by *Phytophthora*.

Researchers have also developed numerous molecular techniques for *P. ramorum* identification.

These include amplifying DNA sequences in the internal transcribed spacer region of the *P. ramorum* genome (ITS polymerase chain reaction, or ITS PCR); real-time PCR, in which DNA abundance is measured in real time during the PCR reaction, using dyes or probes such as SBYR-Green or TaqMan; multiplex PCR, which amplifies more than one region of DNA at the same time; and single strand conformation polymorphism (SSCP), which uses the ITS DNA sequence amplified by the PCR reaction to differentiate *Phytophthora* species according to their differential movement through a gel.

Additionally, researchers have begun using features of the DNA sequence of *P. ramorum* to pinpoint the minuscule differences of separate *P. ramorum* isolates from each other. Two techniques for doing this are amplified fragment length polymorphism, which through comparing differences between various fragments in the sequence has enabled researchers to differentiate correctly between EU and U.S. isolates, and the examination of microsatellites, which are areas on the sequence featuring repeating base pairs. When *P. ramorum* propagules arrive in a new geographic location and establish colonies, these microsatellites begin to display mutation in a relatively short time, and they mutate in a stepwise fashion. Based on this, researchers in California have been able to construct trees, based on microsatellite analyses of isolates collected from around the state, that trace the movement of *P. ramorum* from two likely initial points of establishment in Marin and Santa Cruz Counties and out to subsequent points.

Early detection of *P. ramorum* on a landscape scale begins with the observation of symptoms on individual plants (and/or detecting *P. ramorum* propagules in watercourses. System-atic ground-based monitoring has been difficult within the range of *P. ramorum* because most infected trees stand on a complex mosaic of lands with various ownerships. In some areas, targeted ground-based surveys have been conducted in areas of heavy recreation or visitor use such as parks, trailheads, and boat ramps. In California, when conducting ground-based detection, looking for symptoms on bay laurel is the most effective strategy, since *P. ramorum* infection of true oaks and tanoaks is almost always highly associated with bay laurel, the main epidemiological springboard for the pathogen. Moreover, on many sites in California (though not all), *P. ramorum* can typically be detected from infected bay laurel tissues via culturing techniques year-round; this is not the case for most other hosts, nor is it the case in Oregon, where tanoak is the most reliable host.

As part of a nationwide USDA program, a ground-based detection survey was implemented from 2003 to 2006 in 39 U.S. states to determine whether the pathogen was established outside the West Coast areas already known to be infested. Sampling areas were stratified by environmental variables likely to be conducive to pathogen growth and by proximity to possible points of inoculum introduction such as nurseries. Samples were collected along transects established in potentially susceptible forests or outside the perimeters of nurseries. The only positive samples were collected in California, confirming that *P. ramorum* was not yet established in the environment outside the West Coast.

Aerial surveying has proven useful for detection of *P. ramorum* infestations across large landscapes, although it is not as "early" a technique as some others because it depends on spotting dead tanoak crowns from fixed-wing aircraft. Sophisticated GPS and sketch-mapping technology enable spotters to mark the locations of dead trees so that ground crews can return to the area to sample from nearby vegetation.

Detection of *P. ramorum* in watercourses has emerged as the earliest of early detection methods. This technique employs pear or rhododendron baits suspended in the watercourse using ropes, buckets, mesh bags, or other similar devices. If plants in the watershed are infected with *P. ramorum*, zoospores of the pathogen (as well as other *Phytophthora* spp.) are likely present in adjacent waterways. Under conducive weather conditions, the zoospores are attracted to the baits and infect them, causing lesions that can be isolated to culture the pathogen or analyzed via PCR assay. This method has detected *P. ramorum* at scales ranging from small, intermittent seasonal drainages to the Garcia, Chetco, and South Fork Eel Rivers in California and Oregon (144, 352, and 689 mi2 drainage areas, respectively). It can detect the existence of infected plants in watersheds before any mortality from the infections becomes evident. Of course, it cannot detect the exact locations of those infected plants: at the first sign of *P. ramorum* propagules in the stream, crews must scour the watershed using all available means to find symptomatic vegetation.

A less technical means of detecting *P. ramorum* at the landscape level involves engaging local landowners across the landscape in the search. Many local county agriculture departments and University of California Cooperative Extension offices in California have been able to keep track of the distribution of the pathogen in their regions through reports and samples brought to them by the public. In 2008, the Garbelotto Laboratory at University of California, Berkeley, along with local collaborators, hosted a series of educational events, called "SOD Blitzes", designed to give local landowners basic information about *P. ramorum* and how to identify its symptoms; each participant was provided with a sampling kit, sampled a certain number of trees on his or her property, and returned the samples to the lab for analysis. This kind of citizen science hopefully can help generate an improved map of *P. ramorum* distribution in the areas where the workshops are held.

## Wildland Management

The course that *P. ramorum* management should take depends on a number of factors, including the scale of the landscape upon which one hopes to manage it. Management of *P. ramorum* has been undertaken at the landscape/ regional level in Oregon in the form of a campaign to completely eradicate the pathogen from the forests in which it has been found (mostly private, but also USDA Forest Service and USDI Bureau of Land Management ownership). The eradication campaign involves vigorous early detection by airplane and watercourse monitoring, a U.S. Department of Agriculture Animal and Plant Health Inspection Service (USDA APHIS) and Oregon Department of Agriculture-led quarantine to prevent movement of host materials out of the area where infected trees are found, and immediate removal of *P. ramorum* host vegetation, symptomatic or not, within a 300-foot (91 m) buffer around each infected tree.

The Oregon eradication effort, which began near the town of Brookings in southwest Oregon in 2001, has adapted its management efforts over the years in response to new information about *P. ramorum*. For example, after inoculation trials of various tree species more clearly delineated which hosts are susceptible, the Oregon cooperators began leaving nonhost species such as Douglas fir and red alder on site. In another example, after finding that a small percentage of tanoak stumps that were resprouting on the host removal sites were infected with the pathogen—whether these infections were systemic or reached the sprouts from the surrounding environment is unknown—the cooperators began pretreating trees with very small, targeted amounts of herbicide to kill the root systems of infected tanoaks before cutting them down. The effort has been successful in that while it has not yet completely eradicated the pathogen from Oregon forests, the epidemic in Oregon has not taken the explosive course that it has in California forests.

California, though, faces significant obstacles that preclude it from mounting the same kind of eradication effort. For one thing, the organism was too well established in forests in the Santa Cruz and San Francisco Bay areas by the time the cause of sudden oak death was discovered to enable any eradication effort to succeed. Even in still relatively uninfested areas of the north coast and southern Big Sur, regionally coordinated efforts to manage the pathogen face huge challenges of leadership, coordination, and funding. Nevertheless, land managers are still working to coordinate efforts between states, counties, and agencies to provide *P. ramorum* management in a more comprehensive manner.

Several options exist for landowners who want to limit the impacts of SOD death on their properties. None of these options is foolproof, guaranteed to eradicate *P. ramorum*, or guaranteed to prevent a tree from becoming infected. Some are still in the initial stage of testing. Nevertheless, when used thoughtfully and thoroughly, some of the treatments do improve the likelihood of either slowing the spread of the pathogen or of limiting its impacts on trees or stands of trees. Assuming that the landowner has correctly identified the host tree(s) and symptom(s), has submitted a sample to a local authority to send to an approved laboratory for testing, and has received confirmation that the tree(s) are indeed infected with *P. ramorum*—or, alternatively, assuming that the landowner knows that *P. ramorum*-infected trees are nearby and wants to protect the resources on his or her property—he or she can attempt control by cultural (individual-tree), chemical, or silvicultural (stand-level) means.

The best evidence that cultural techniques might help protect trees against *P. ramorum* comes from research that has established a correlation between disease risk in coast live oak trees and the trees' proximity to bay laurel. In particular, this research found that bay laurel trees growing within 5 m of the trunk of an oak tree were the best predictors of disease risk. This implies that strategic removal of bay laurel trees near coast live oaks might decrease the risk of oak infection. Wholesale removal of bay laurel trees would not be warranted, since the bay laurels close to the oak trees appear to provide the greatest risk factor. Whether the same pattern is true for other oaks or tanoaks has yet to be established. Research on this subject has been started for tanoak, but any eventual cultural recommendations will be more complicated, because tanoak twigs also serve as sources of *P. ramorum* inoculum.

A promising treatment for preventing infection of individual oak and tanoak trees—not for curing an already established infection—is a phosphonate fungicide marketed under the trade name Agri-fos. Phosphonate is a neutralized form of phosphorous acid that works not by direct antagonism of *Phytophthora*, but by stimulating various kinds of immune responses on the part of the tree. It is mostly environmentally benign if not applied to nontarget plants and can be applied either as an injection into the tree stem or as a spray to the bole. When applying Agri-fos as a spray, it must be combined with an organosilicate surfactant, Pentra-bark, to enable the product to adhere to the bole long enough to be absorbed by the tree. Agri-fos has been very effective in preventing tree infections, but it must be applied when visible symptoms of *P. ramorum* on other trees in the immediate neighborhood are still relatively distant; otherwise, the tree to be treated likely is already infected, but visible symptoms have not yet developed (especially true for tanoak).

Trials of silvicultural methods for treating *P. ramorum* began in Humboldt County in northwest coastal California in 2006. The trials have taken place on a variety of infested properties both private and public and have generally focused on varying levels and kinds of host removal. The largest (50 acres (200,000 m²)) and most replicated trials have involved removal of tanoak and bay laurel by chainsaw throughout the infested stand, both with and without subsequent under-burning designed to eliminate small seedlings and infested leaf litter. Other treatments included host removal in a modified "shaded fuelbreak" design in which all bay laurel is removed, but not all tanoaks; bay and tanoak removal using herbicides; and removal of bay laurel alone. The results of these treatments are still being monitored, but repeated sampling has so far detected only very small amounts of *P. ramorum* in the soil or on vegetation in the treated sites.

## Nursery Management

Research and development in managing *P. ramorum* in nursery settings extends from *P. ramorum* in the individual plant, to *P. ramorum* in the nursery environment, to the pathogen's movement across state and national borders in infected plants.

An array of studies have tested the curative and protective effects of various chemical compounds against *P. ramorum* in plants valued as ornamentals or Christmas trees. Many studies have focused on the four main ornamental hosts of *P. ramorum* (*Rhododendron*, *Camellia*, *Viburnum*, and *Pieris*). Several effective compounds have been found; some of the most effective include mefenoxam, metalaxyl, dimethomorph, and fenamidone. Many of these studies have converged upon the following conclusions: chemical compounds are, in general, more effective as preventives than in curatives; when used preventively, chemical compounds must be reapplied at various intervals; and chemical compounds can mask the symptoms of *P. ramorum* infection in the host plant, potentially interfering with inspections for quarantine efforts. In general, these compounds suppress but do not eradicate the pathogen, and some researchers are concerned that with repeated use the pathogen may become resistant to them. These studies and conclusions are summarized by Kliejunas.

Another area of research and evolving practice deals with eliminating *P. ramorum* from nursery environments in which it is established to prevent human-mediated pathogen movement within the ornamental plant trade. One way of approaching this is through a robust quarantine and inspection program, which the various federal and state regulatory agencies have implemented. Under the federal *P. ramorum* quarantine program implemented by USDA APHIS, nurseries in California, Oregon, and Washington are regulated and must participate in an annual inspection regimen; nurseries in the 14 infested counties in coastal California, plus the limited infested area in Curry County, Oregon, must participate in a more stringent inspection schedule when shipping out of this area.

Much of the research into disinfesting nurseries has focused on the voluntary best management practices (BMPs) that nurseries can implement to prevent *P. ramorum*'s introduction into the nursery and movement from plant to plant. In 2008, a group of nursery industry organizations issued a list of BMPs that includes subsections on pest prevention/management, training, internal/external monitoring/audits, records/traceability, and documentation. The document includes such specific recommendations as "Avoid overhead irrigation of high-risk plants"; "After every crop rotation, disinfect propagation mist beds, sorting area, cutting benches, machines and tools to minimize the spread or introduction of pathogens"; and "Nursery personnel should attend one or more *P. ramorum* trainings conducted by qualified personnel or document self-training".

Research on control of *P. ramorum* in nurseries has also focused on disinfesting irrigation water containing *P. ramorum* inoculum. Irrigation water can become infested from bay trees in the forest (if the irrigation source is a stream), from bay trees overhanging irrigation ponds, from runoff from infested forests, or from recirculated irrigation water. Experiments in Germany with three types of filters—slow sand filters, lava filters, and constructed wetlands—showed that the first two removed *P. ramorum* from the irrigation water completely, while 37% of the post-treatment water samples from the constructed wetland still contained *P. ramorum*.

Since *P. ramorum* can persist for an undetermined period of time within the soil profile, management programs in nurseries should also deal with delineating the pathogen's distribution in nursery soil and eliminating it from infested areas. A variety of chemical options has been tested for soil disinfestation, including such chemicals as chloropicrin, metham sodium, iodomethane, and dazomet. Lab tests indicated that all of these chemicals were effective when applied to infested soil in glass jars. Additionally, tests on volunteer nurseries with infested soil demonstrated that dazomet (trade name Basamid) fumigation followed by a 14-day tarping period successfully removed *P. ramorum* from the soil profile. Other soil disinfestation practices under investigation, or in which interest has been expressed, include steam sterilization, solarization, and paving of infested areas.

## General Sanitation in Infested Areas

One of the most important aspects of *P. ramorum* control involves interrupting the human-mediated movement of the pathogen by ensuring that infested materials do not move from location to location. While enforceable quarantines perform part of this function, basic cleanliness when working or recreating in infested areas is also important. In most cases, cleanliness practices involve ridding potentially infested surfaces—such as shoes, vehicles, and pets—of foliage and mud before leaving the infested area. The demands of implementing these practices become more complex when large numbers of people are working in infested areas, as in construction, timber harvesting, or wildfire suppression. The California Department of Forestry and Fire Protection and USDA Forest Service have implemented guidelines and mitigation requirements for the latter two situations; basic information about cleanliness in *P. ramorum*-infested areas can be found at the California Oak Mortality Task Force web site (www.suddenoakdeath.org) under the "Treatment and Management" section (subsection "Sanitation and Reducing Spread").

## Government Agency Involvement

In England in 2009, the Forestry Commission, DEFRA, the Food and Environment Research Agency, Cornwall County Council, and Natural England are working together to record the locations and deal with this disease. Natural England is offering grant funding through its Environmental Stewardship, Countryside Stewardship, and Environmentally Sensitive Area schemes to clear rhododendron. In 2011, the Forestry Commission started felling 10,000 acres (40 km²) of larch forest in the south-west of England, as an attempt to halt the spread of the disease. In Northern Ireland at the end of 2011, the Department of Agriculture and Rural Development's Forest Service began felling 14 hectares of affected Larch woodland at Moneyscalp, on the edge of Tollymore Forest Park in County Down.

# Bacteria

Bacteria constitute a large domain of prokaryotic microorganisms. Typically a few micrometres in length, bacteria have a number of shapes, ranging from spheres to rods and spirals. Bacteria were among the first life forms to appear on Earth, and are present in most of its habitats. Bacteria inhabit soil, water, acidic hot springs, radioactive waste, and the deep portions of Earth's crust. Bacteria also live in symbiotic and parasitic relationships with plants and animals.

There are typically 40 million bacterial cells in a gram of soil and a million bacterial cells in a millilitre of fresh water. There are approximately $5 \times 10^{30}$ bacteria on Earth, forming a biomass which exceeds that of all plants and animals. Bacteria are vital in recycling nutrients, with many of the stages in nutrient cycles dependent on these organisms, such as the fixation of nitrogen from the atmosphere and putrefaction. In the biological communities surrounding hydrothermal vents and cold seeps, bacteria provide the nutrients needed to sustain life by converting dissolved compounds, such as hydrogen sulphide and methane, to energy. On 17 March 2013, researchers reported data that suggested bacterial life forms thrive in the Mariana Trench, which with a depth of up to 11 kilometres is the deepest part of the Earth's oceans. Other researchers reported related studies that microbes thrive inside rocks up to 580 metres below the sea floor under 2.6 kilometres of ocean off the coast of the northwestern United States. According to one of the researchers, "You can find microbes everywhere — they're extremely adaptable to conditions, and survive wherever they are."

Most bacteria have not been characterized, and only about half of the bacterial phyla have species that can be grown in the laboratory. The study of bacteria is known as bacteriology, a branch of microbiology.

There are approximately ten times as many bacterial cells in the human flora as there are human cells in the body, with the largest number of the human flora being in the gut flora, and a large number on the skin. The vast majority of the bacteria in the body are rendered harmless by the protective effects of the immune system, and some are beneficial. However, several species of bacteria are pathogenic and cause infectious diseases, including cholera, syphilis, anthrax, leprosy, and bubonic plague. The most common fatal bacterial diseases are respiratory infections, with tuberculosis alone killing about 2 million people per year, mostly in sub-Saharan Africa. In developed countries, antibiotics are used to treat bacterial infections and are also used in farming, making antibiotic resistance a growing problem. In industry, bacteria are important in sewage treatment and the breakdown of oil spills, the production of cheese and yogurt through fermentation, and the recovery of gold, palladium, copper and other metals in the mining sector, as well as in biotechnology, and the manufacture of antibiotics and other chemicals.

Once regarded as plants constituting the class *Schizomycetes*, bacteria are now classified as prokaryotes. Unlike cells of animals and other eukaryotes, bacterial cells do not contain a nucleus and rarely harbour membrane-bound organelles. Although the term *bacteria* traditionally included all prokaryotes, the scientific classification changed after the discovery in the 1990s that prokaryotes consist of two very different groups of organisms that evolved from an ancient common ancestor. These evolutionary domains are called *Bacteria* and *Archaea*.

## Origin and Early Evolution

The ancestors of modern bacteria were unicellular microorganisms that were the first forms of life to appear on Earth, about 4 billion years ago. For about 3 billion years, most organisms were microscopic, and bacteria and archaea were the dominant forms of life. In 2008, fossils of macroorganisms were discovered and named as the Francevillian biota. Although bacterial fossils exist, such as stromatolites, their lack of distinctive morphology prevents them from being used to examine the history of bacterial evolution, or to date the time of origin of a particular bacterial species. However, gene sequences can be used to reconstruct the bacterial phylogeny, and these studies indicate that bacteria diverged first from the archaeal/eukaryotic lineage. Bacteria were also involved in the second great evolutionary divergence, that of the archaea and eukaryotes. Here, eukaryotes resulted from the entering of ancient bacteria into endosymbiotic associations with the ancestors of eukaryotic cells, which were themselves possibly related to the Archaea. This involved the engulfment by proto-eukaryotic cells of alphaproteobacterial symbionts to form either mitochondria or hydrogenosomes, which are still found in all known Eukarya (sometimes in highly reduced form, e.g. in ancient "amitochondrial" protozoa). Later on, some eukaryotes that already contained mitochondria also engulfed cyanobacterial-like organisms. This led to the formation of chloroplasts in algae and plants. There are also some algae that originated from even later endosymbiotic events. Here, eukaryotes engulfed a eukaryotic algae that developed into a "second-generation" plastid. This is known as secondary endosymbiosis.

## Morphology

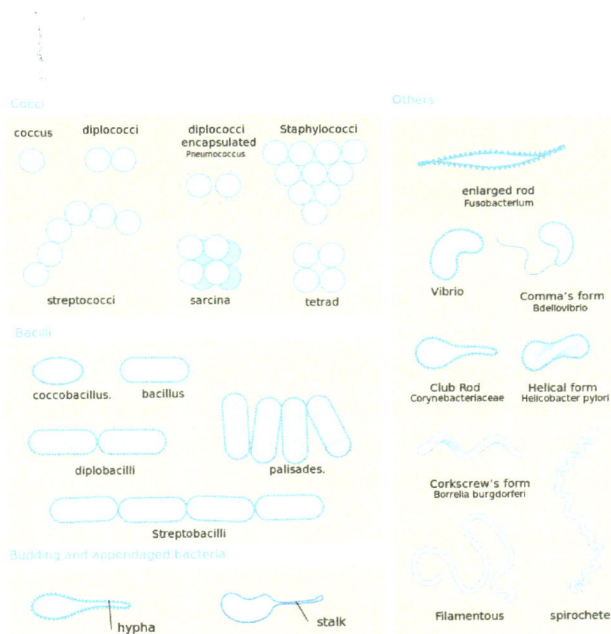

Bacteria display many cell morphologies and arrangements

Bacteria display a wide diversity of shapes and sizes, called *morphologies*. Bacterial cells are about one-tenth the size of eukaryotic cells and are typically 0.5–5.0 micrometres in length. However, a few species are visible to the unaided eye — for example, *Thiomargarita namibiensis* is up to half a millimetre long and *Epulopiscium fishelsoni* reaches 0.7 mm. Among the smallest bacteria are members of the genus *Mycoplasma*, which measure only 0.3 micrometres, as small as the largest viruses. Some bacteria may be even smaller, but these ultramicrobacteria are not well-studied.

Most bacterial species are either spherical, called *cocci* (*sing*. coccus, from Greek *kókkos*, grain, seed), or rod-shaped, called *bacilli* (*sing*. bacillus, from Latin *baculus*, stick). Elongation is associated with swimming. Some bacteria, called *vibrio*, are shaped like slightly curved rods or comma-shaped; others can be spiral-shaped, called *spirilla*, or tightly coiled, called *spirochaetes*. A small number of species even have tetrahedral or cuboidal shapes. More recently, some bacteria were discovered deep under Earth's crust that grow as branching filamentous types with a star-shaped cross-section. The large surface area to volume ratio of this morphology may give these bacteria an advantage in nutrient-poor environments. This wide variety of shapes is determined by the bacterial cell wall and cytoskeleton, and is important because it can influence the ability of bacteria to acquire nutrients, attach to surfaces, swim through liquids and escape predators.

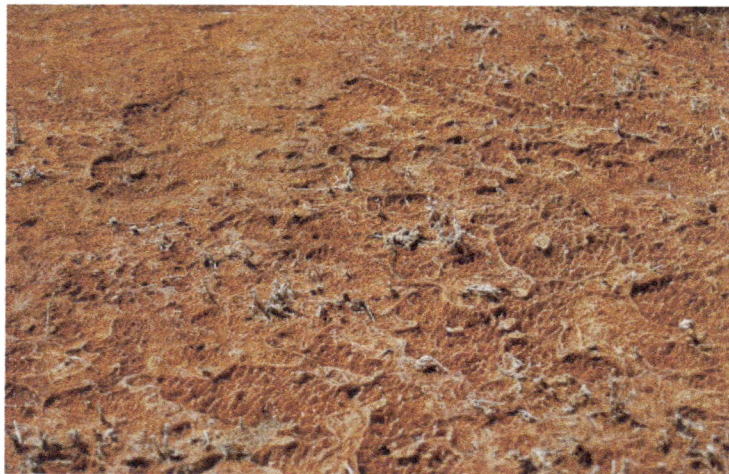

A biofilm of thermophilic bacteria in the outflow of Mickey Hot Springs, Oregon, approximately 20 mm thick.

Many bacterial species exist simply as single cells, others associate in characteristic patterns: *Neisseria* form diploids (pairs), *Streptococcus* form chains, and *Staphylococcus* group together in "bunch of grapes" clusters. Bacteria can also be elongated to form filaments, for example the Actinobacteria. Filamentous bacteria are often surrounded by a sheath that contains many individual cells. Certain types, such as species of the genus *Nocardia*, even form complex, branched filaments, similar in appearance to fungal mycelia.

Bacteria often attach to surfaces and form dense aggregations called *biofilms* or bacterial mats. These films can range from a few micrometers in thickness to up to half a meter in depth, and may contain multiple species of bacteria, protists and archaea. Bacteria living in biofilms display a complex arrangement of cells and extracellular components, forming secondary structures, such as microcolonies, through which there are networks of channels to enable better diffusion of nutrients. In natural environments, such as soil or the surfaces of plants, the majority of bacteria are bound to surfaces in biofilms. Biofilms are also important in medicine, as these structures are often present during chronic bacterial infections or in infections of implanted medical devices, and bacteria protected within biofilms are much harder to kill than individual isolated bacteria.

Even more complex morphological changes are sometimes possible. For example, when starved of amino acids, Myxobacteria detect surrounding cells in a process known as quorum sensing, migrate toward each other, and aggregate to form fruiting bodies up to 500 micrometres long and containing approximately 100,000 bacterial cells. In these fruiting bodies, the bacteria perform separate tasks; this type of cooperation is a simple type of multicellular organisation. For example, about one in 10 cells migrate to the top of these fruiting bodies and differentiate into a specialised dormant state called myxospores, which are more resistant to drying and other adverse environmental conditions than are ordinary cells.

## Cellular Structure

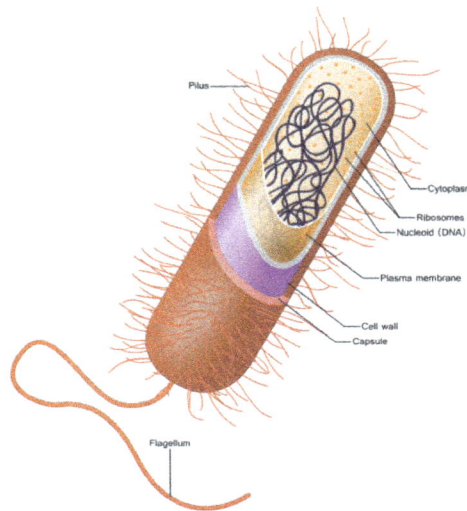

Structure and contents of a typical gram-positive bacterial cell (seen by the fact that only *one* cell membrane is present).

## Intracellular Structures

The bacterial cell is surrounded by a cell membrane (also known as a lipid, cytoplasmic or plasma membrane). This membrane encloses the contents of the cell and acts as a barrier to hold nutrients, proteins and other essential components of the *cytoplasm* within the cell. As they are prokaryotes, bacteria do not usually have membrane-bound organelles in their cytoplasm, and thus contain few large intracellular structures. They lack a true nucleus, mitochondria, chloroplasts and the other organelles present in eukaryotic cells. Bacteria were once seen as simple bags of cytoplasm, but structures such as the *prokaryotic cytoskeleton* and the localization of proteins to specific locations within the cytoplasm that give bacteria some complexity have been discovered. These subcellular levels of organization have been called "bacterial hyperstructures".

*Bacterial microcompartments*, such as carboxysomes, provide a further level of organization; they are compartments within bacteria that are surrounded by polyhedral protein shells, rather than by lipid membranes. These "polyhedral organelles" localize and compartmentalize bacterial metabolism, a function performed by the membrane-bound organelles in eukaryotes.

Many important biochemical reactions, such as energy generation, use concentration gradients across membranes. The general lack of internal membranes in bacteria means reactions such as electron transport occur across the cell membrane between the cytoplasm and the *periplasmic space*. However, in many photosynthetic bacteria the plasma membrane is highly folded and fills most of the cell with layers of light-gathering membrane. These light-gathering complexes may even form lipid-enclosed structures called chlorosomes in green sulfur bacteria. Other proteins import nutrients across the cell membrane, or expel undesired molecules from the cytoplasm.

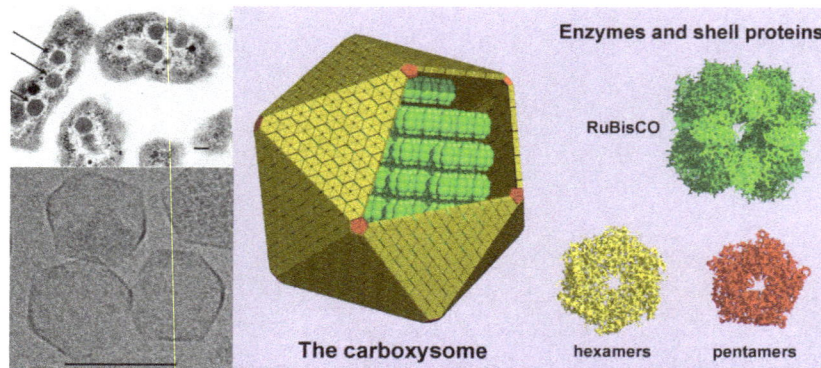

Carboxysomes are protein-enclosed bacterial organelles. Top left is an electron microscope image of carboxysomes in *Halothiobacillus neapolitanus*, below is an image of purified carboxysomes. On the right is a model of their structure. Scale bars are 100 nm.

Bacteria do not have a membrane-bound nucleus, and their genetic material is typically a single circular DNA chromosome located in the cytoplasm in an irregularly shaped body called the *nucleoid*. The nucleoid contains the chromosome with its associated proteins and RNA. The phylum Planctomycetes and candidate phylum Poribacteria may be exceptions to the general absence of internal membranes in bacteria, because they appear to have a double membrane around their nucleoids and contain other membrane-bound cellular structures. Like all living organisms, bacteria contain *ribosomes*, often grouped in chains called polyribosomes, for the production of proteins, but the structure of the bacterial ribosome is different from that of eukaryotes and Archaea. Bacterial ribosomes have a sedimentation rate of 70S (measured in Svedberg units): their subunits have rates of 30S and 50S. Some antibiotics bind specifically to 70S ribosomes and inhibit bacterial protein synthesis. Those antibiotics kill bacteria without affecting the larger 80S ribosomes of eukaryotic cells and without harming the host.

Some bacteria produce intracellular nutrient storage granules for later use, such as glycogen, polyphosphate, sulfur or polyhydroxyalkanoates. Certain bacterial species, such as the photosynthetic Cyanobacteria, produce internal gas vesicles, which they use to regulate their buoyancy – allowing them to move up or down into water layers with different light intensities and nutrient levels. *Intracellular membranes* called *chromatophores* are also found in membranes of phototrophic bacteria. Used primarily for photosynthesis, they contain bacteriochlorophyll pigments and carotenoids. An early idea was that bacteria might contain membrane folds termed mesosomes, but these were later shown to be artifacts produced by the chemicals used to prepare the cells for electron microscopy. *Inclusions* are considered to be nonliving components of the cell that do not possess metabolic activity and are not bounded by membranes. The most common inclusions are glycogen, lipid droplets, crystals, and pigments. *Volutin granules* are cytoplasmic inclusions of complexed inorganic polyphosphate. These granules are called *metachromatic granules* due to their displaying the metachromatic effect; they appear red or blue when stained with the blue dyes methylene blue or toluidine blue. *Gas vacuoles*, which are freely permeable to gas, are membrane-bound vesicles present in some species of *Cyanobacteria*. They allow the bacteria to control their buoyancy. *Microcompartments* are widespread, membrane-bound organelles that are made of a protein shell that surrounds and encloses various enzymes. *Carboxysomes* are bacterial microcompartments that contain enzymes involved in carbon fixation. *Magnetosomes* are bacterial microcompartments, present in magnetotactic bacteria, that contain magnetic crystals.

## Extracellular Structures

In most bacteria, a *cell wall* is present on the outside of the cell membrane. The cell membrane and cell wall comprise the *cell envelope*. A common bacterial cell wall material is *peptidoglycan* (called "murein" in older sources), which is made from polysaccharide chains cross-linked by peptides containing D-amino acids. Bacterial cell walls are different from the cell walls of plants and fungi, which are made of cellulose and chitin, respectively. The cell wall of bacteria is also distinct from that of Archaea, which do not contain peptidoglycan. The cell wall is essential to the survival of many bacteria, and the antibiotic penicillin is able to kill bacteria by inhibiting a step in the synthesis of peptidoglycan.

There are broadly speaking two different types of cell wall in bacteria, a thick one in the gram-positives and a thinner one in the gram-negatives. The names originate from the reaction of cells to the Gram stain, a test long-employed for the classification of bacterial species.

*Gram-positive bacteria* possess a thick cell wall containing many layers of peptidoglycan and *teichoic acids*. In contrast, *gram-negative bacteria* have a relatively thin cell wall consisting of a few layers of peptidoglycan surrounded by a second lipid membrane containing *lipopolysaccharides* and lipoproteins. Lipopolysaccharides, also called *endotoxins*, are composed of polysaccharides and *lipid A* that is responsible for much of the toxicity of gram-negative bacteria. Most bacteria have the gram-negative cell wall, and only the Firmicutes and Actinobacteria have the alternative gram-positive arrangement. These two groups were previously known as the low G+C and high G+C gram-positive bacteria, respectively. These differences in structure can produce differences in antibiotic susceptibility; for instance, vancomycin can kill only gram-positive bacteria and is ineffective against gram-negative pathogens, such as *Haemophilus influenzae* or *Pseudomonas aeruginosa*. If the bacterial cell wall is entirely removed, it is called a *protoplast*, whereas if it is partially removed, it is called a *spheroplast*. β-Lactam antibiotics, such as penicillin, inhibit the formation of peptidoglycan cross-links in the bacterial cell wall. The enzyme lysozyme, found in human tears, also digests the cell wall of bacteria and is the body's main defense against eye infections.

*Acid-fast bacteria*, such as *Mycobacteria*, are resistant to decolorization by acids during staining procedures. The high mycolic acid content of *Mycobacteria*, is responsible for the staining pattern of poor absorption followed by high retention. The most common staining technique used to identify acid-fast bacteria is the Ziehl-Neelsen stain or acid-fast stain, in which the acid-fast bacilli are stained bright-red and stand out clearly against a blue background. *L-form bacteria* are strains of bacteria that lack cell walls. The main pathogenic bacteria in this class is *Mycoplasma* (not to be confused with *Mycobacteria*).

In many bacteria, an *S-layer* of rigidly arrayed protein molecules covers the outside of the cell. This layer provides chemical and physical protection for the cell surface and can act as a macro-molecular diffusion barrier. S-layers have diverse but mostly poorly understood functions, but are known to act as virulence factors in *Campylobacter* and contain surface enzymes in *Bacillus stearothermophilus*.

*Flagella* are rigid protein structures, about 20 nanometres in diameter and up to 20 micrometres in length, that are used for motility. Flagella are driven by the energy released by the transfer of ions down an electrochemical gradient across the cell membrane.

*Fimbriae* (sometimes called "attachment pili") are fine filaments of protein, usually 2–10 nano-metres in diameter and up to several micrometers in length. They are distributed over the sur-face of the cell, and resemble fine hairs when seen under the electron microscope. Fimbriae are believed to be involved in attachment to solid surfaces or to other cells, and are essential for the virulence of some bacterial pathogens. *Pili* (*sing.* pilus) are cellular appendages, slight-ly larger than fimbriae, that can transfer genetic material between bacterial cells in a process called conjugation where they are called *conjugation pili* or "sex pili". They can also generate movement where they are called *type IV pili*.

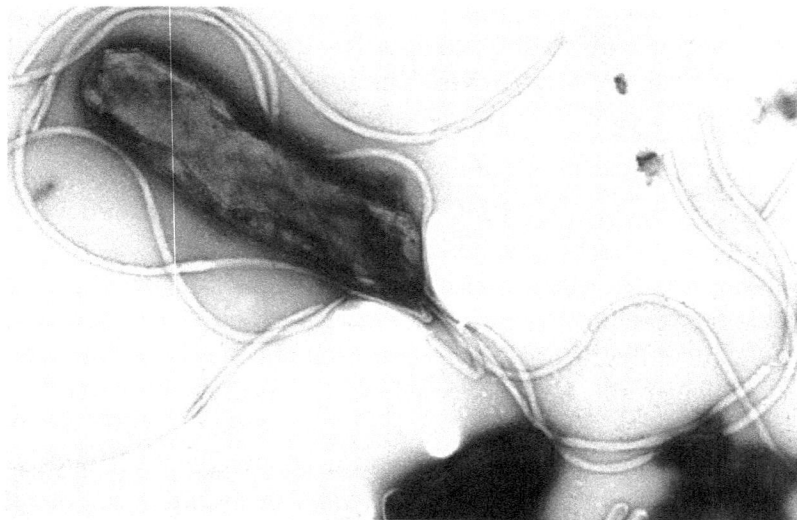

*Helicobacter pylori* electron micrograph, showing multiple flagella on the cell surface

*Glycocalyx* are produced by many bacteria to surround their cells, and vary in structural complex-ity: ranging from a disorganised *slime layer* of extra-cellular polymer to a highly structured *cap-sule*. These structures can protect cells from engulfment by eukaryotic cells such as macrophages (part of the human immune system). They can also act as antigens and be involved in cell recogni-tion, as well as aiding attachment to surfaces and the formation of biofilms.

The assembly of these extracellular structures is dependent on bacterial secretion systems. These transfer proteins from the cytoplasm into the periplasm or into the environment around the cell. Many types of secretion systems are known and these structures are often essential for the viru-lence of pathogens, so are intensively studied.

# Endospores

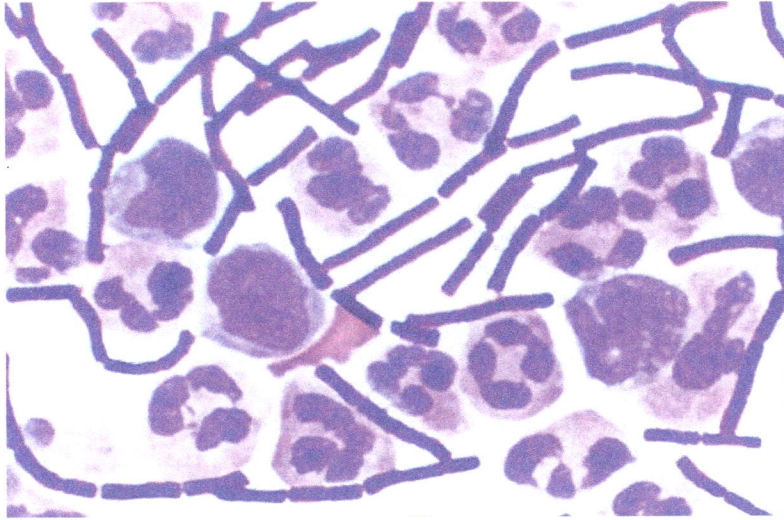

*Bacillus anthracis* (stained purple) growing in cerebrospinal fluid

Certain genera of gram-positive bacteria, such as *Bacillus, Clostridium, Sporohalobacter, Anaerobacter*, and *Heliobacterium*, can form highly resistant, dormant structures called *endospores*. In almost all cases, one endospore is formed and this is not a reproductive process, although *Anaerobacter* can make up to seven endospores in a single cell. Endospores have a central core of cytoplasm containing DNA and ribosomes surrounded by a cortex layer and protected by an impermeable and rigid coat. Dipicolinic acid is a chemical compound that composes 5% to 15% of the dry weight of bacterial spores. It is implicated as responsible for the heat resistance of the endospore.

Endospores show no detectable metabolism and can survive extreme physical and chemical stresses, such as high levels of UV light, gamma radiation, detergents, disinfectants, heat, freezing, pressure, and desiccation. In this dormant state, these organisms may remain viable for millions of years, and endospores even allow bacteria to survive exposure to the vacuum and radiation in space. According to scientist Dr. Steinn Sigurdsson, "There are viable bacterial spores that have been found that are 40 million years old on Earth — and we know they're very hardened to radiation." Endospore-forming bacteria can also cause disease: for example, anthrax can be contracted by the inhalation of *Bacillus anthracis* endospores, and contamination of deep puncture wounds with *Clostridium tetani* endospores causes tetanus.

## Metabolism

Bacteria exhibit an extremely wide variety of metabolic types. The distribution of metabolic traits within a group of bacteria has traditionally been used to define their taxonomy, but these traits often do not correspond with modern genetic classifications. Bacterial metabolism is classified into nutritional groups on the basis of three major criteria: the kind of energy used for growth, the source of carbon, and the electron donors used for growth. An additional criterion of respiratory microorganisms are the electron acceptors used for aerobic or anaerobic respiration.

| Nutritional types in bacterial metabolism | | | |
|---|---|---|---|
| **Nutritional type** | **Source of energy** | **Source of carbon** | **Examples** |
| Phototrophs | Sunlight | Organic compounds (photoheterotrophs) or carbon fixation (photoautotrophs) | Cyanobacteria, Green sulfur bacteria, Chloroflexi, or Purple bacteria |
| Lithotrophs | Inorganic compounds | Organic compounds (lithoheterotrophs) or carbon fixation (lithoautotrophs) | Thermodesulfobacteria, *Hydrogenophilaceae*, or Nitrospirae |
| Organotrophs | Organic compounds | Organic compounds (chemoheterotrophs) or carbon fixation (chemoautotrophs) | *Bacillus, Clostridium* or *Enterobacteriaceae* |

Carbon metabolism in bacteria is either *heterotrophic*, where organic carbon compounds are used as carbon sources, or *autotrophic*, meaning that cellular carbon is obtained by fixing carbon dioxide. Heterotrophic bacteria include parasitic types. Typical autotrophic bacteria are phototrophic cyanobacteria, green sulfur-bacteria and some purple bacteria, but also many chemolithotrophic species, such as nitrifying or sulfur-oxidising bacteria. Energy metabolism of bacteria is either based on *phototrophy*, the use of light through photosynthesis, or based on *chemotrophy*, the use of chemical substances for energy, which are mostly oxidised at the expense of oxygen or alternative electron acceptors (aerobic/anaerobic respiration).

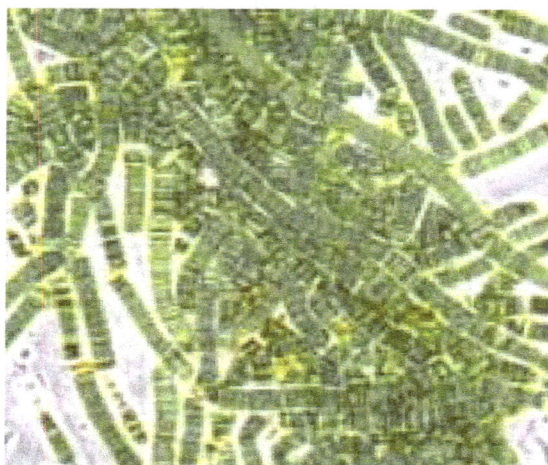

Filaments of photosynthetic cyanobacteria

Bacteria are further divided into *lithotrophs* that use inorganic electron donors and *organotrophs* that use organic compounds as electron donors. Chemotrophic organisms use the respective electron donors for energy conservation (by aerobic/anaerobic respiration or fermentation) and biosynthetic reactions (e.g., carbon dioxide fixation), whereas phototrophic organisms use them only for biosynthetic purposes. Respiratory organisms use chemical compounds as a source of energy by taking electrons from the reduced substrate and transferring them to a terminal electron acceptor in a redox reaction. This reaction releases energy that can be used to synthesise ATP and drive metabolism. In *aerobic organisms*, oxygen is used as the electron acceptor. In *anaerobic organisms* other inorganic compounds, such as nitrate, sulfate or carbon dioxide are used as electron acceptors. This leads to the ecologically important processes of denitrification, sulfate reduction, and acetogenesis, respectively.

Another way of life of chemotrophs in the absence of possible electron acceptors is fermentation, wherein the electrons taken from the reduced substrates are transferred to oxidised intermediates to generate reduced fermentation products (e.g., lactate, ethanol, hydrogen, butyric acid). Fermentation is possible, because the energy content of the substrates is higher than that of the products, which allows the organisms to synthesise ATP and drive their metabolism.

These processes are also important in biological responses to pollution; for example, sulfate-reducing bacteria are largely responsible for the production of the highly toxic forms of mercury (methyl- and dimethylmercury) in the environment. Non-respiratory anaerobes use fermentation to generate energy and reducing power, secreting metabolic by-products (such as ethanol in brewing) as waste. Facultative anaerobes can switch between fermentation and different terminal electron acceptors depending on the environmental conditions in which they find themselves.

Lithotrophic bacteria can use inorganic compounds as a source of energy. Common inorganic electron donors are hydrogen, carbon monoxide, ammonia (leading to nitrification), ferrous iron and other reduced metal ions, and several reduced sulfur compounds. In unusual circumstances, the gas methane can be used by methanotrophic bacteria as both a source of electrons and a substrate for carbon anabolism. In both aerobic phototrophy and chemolithotrophy, oxygen is used as a terminal electron acceptor, whereas under anaerobic conditions inorganic compounds are used instead. Most lithotrophic organisms are autotrophic, whereas organotrophic organisms are heterotrophic.

In addition to fixing carbon dioxide in photosynthesis, some bacteria also fix nitrogen gas (nitrogen fixation) using the enzyme nitrogenase. This environmentally important trait can be found in bacteria of nearly all the metabolic types listed above, but is not universal.

Regardless of the type of metabolic process they employ, the majority of bacteria are able to take in raw materials only in the form of relatively small molecules, which enter the cell by diffusion or through molecular channels in cell membranes. The Planctomycetes are the exception (as they are in possessing membranes around their nuclear material). It has recently been shown that *Gemmata obscuriglobus* is able to take in large molecules via a process that in some ways resembles endocytosis, the process used by eukaryotic cells to engulf external items.

## Growth and Reproduction

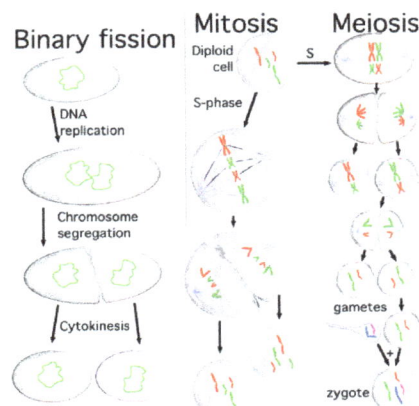

Many bacteria reproduce through binary fission, which is compared to mitosis and meiosis in this image.

Unlike in multicellular organisms, increases in cell size (cell growth) and reproduction by cell division are tightly linked in unicellular organisms. Bacteria grow to a fixed size and then reproduce through *binary fission*, a form of asexual reproduction. Under optimal conditions, bacteria can grow and divide extremely rapidly, and bacterial populations can double as quickly as every 9.8 minutes. In cell division, two identical clone daughter cells are produced. Some bacteria, while still reproducing asexually, form more complex reproductive structures that help disperse the newly formed daughter cells. Examples include fruiting body formation by *Myxobacteria* and aerial hyphae formation by *Streptomyces*, or budding. Budding involves a cell forming a protrusion that breaks away and produces a daughter cell.

A colony of *Escherichia coli*

In the laboratory, bacteria are usually grown using solid or liquid media. Solid *growth media*, such as agar plates, are used to isolate pure cultures of a bacterial strain. However, liquid growth media are used when measurement of growth or large volumes of cells are required. Growth in stirred liquid media occurs as an even cell suspension, making the cultures easy to divide and transfer, although isolating single bacteria from liquid media is difficult. The use of selective media (media with specific nutrients added or deficient, or with antibiotics added) can help identify specific organisms.

Most laboratory techniques for growing bacteria use high levels of nutrients to produce large amounts of cells cheaply and quickly. However, in natural environments, nutrients are limited, meaning that bacteria cannot continue to reproduce indefinitely. This nutrient limitation has led the evolution of different growth strategies (r/K selection theory). Some organisms can grow extremely rapidly when nutrients become available, such as the formation of algal (and cyanobacterial) blooms that often occur in lakes during the summer. Other organisms have adaptations to harsh environments, such as the production of multiple antibiotics by *Streptomyces* that inhibit the growth of competing microorganisms. In nature, many organisms live in communities (e.g., biofilms) that may allow for increased supply of nutrients and protection from environmental stresses. These relationships can be essential for growth of a particular organism or group of organisms (syntrophy).

*Bacterial growth* follows four phases. When a population of bacteria first enter a high-nutrient environment that allows growth, the cells need to adapt to their new environment. The first phase of growth is the *lag phase*, a period of slow growth when the cells are adapting to the high-nutrient environment and preparing for fast growth. The lag phase has high biosynthesis rates, as proteins necessary for rapid growth are produced. The second phase of growth is the *log phase*, also known as the *logarithmic or exponential phase*. The log phase is marked by rapid exponential growth. The rate at which cells grow during this phase is known as the *growth rate* ($k$), and the time it takes the cells to double is known as the *generation time* ($g$). During log phase, nutrients are metabolised at maximum speed until one of the nutrients is depleted and starts limiting growth. The third phase of growth is the *stationary phase* and is caused by depleted nutrients. The cells reduce their metabolic activity and consume non-essential cellular proteins. The stationary phase is a transition from rapid growth to a stress response state and there is increased expression of genes involved in DNA repair, antioxidant metabolism and nutrient transport. The final phase is the *death phase* where the bacteria run out of nutrients and die.

## Genetics

Most bacteria have a single circular chromosome that can range in size from only 160,000 base pairs in the endosymbiotic bacteria *Candidatus Carsonella ruddii*, to 12,200,000 base pairs in the soil-dwelling bacteria *Sorangium cellulosum*. Spirochaetes of the genus *Borrelia* are a notable exception to this arrangement, with bacteria such as *Borrelia burgdorferi*, the cause of Lyme disease, containing a single linear chromosome. The genes in bacterial genomes are usually a single continuous stretch of DNA and although several different types of introns do exist in bacteria, these are much rarer than in eukaryotes.

Bacteria may also contain *plasmids*, which are small extra-chromosomal DNAs that may contain genes for antibiotic resistance or virulence factors. Plasmids replicate independently of chromosomes, so it is possible that plasmids could be lost in bacterial cell division. Against this possibility is the fact that a single bacterium can contain hundreds of copies of a single plasmid.

Bacteria, as asexual organisms, inherit identical copies of their parent's genes (i.e., they are clonal). However, all bacteria can evolve by selection on changes to their genetic material DNA caused by genetic recombination or mutations. Mutations come from errors made during the replication of DNA or from exposure to mutagens. Mutation rates vary widely among different species of bacteria and even among different clones of a single species of bacteria. Genetic changes in bacterial genomes come from either random mutation during replication or "stress-directed mutation", where genes involved in a particular growth-limiting process have an increased mutation rate.

## DNA Transfer

Some bacteria also transfer genetic material between cells. This can occur in three main ways. First, bacteria can take up exogenous DNA from their environment, in a process called *transformation*. Genes can also be transferred by the process of *transduction*, when the integration of a bacteriophage introduces foreign DNA into the chromosome. The third method of gene transfer is *conjugation*, whereby DNA is transferred through direct cell contact.

Transduction of bacterial genes by bacteriophage appears to be a consequence of infrequent errors during intracellular assembly of virus particles, rather than a bacterial adaptation.

Conjugation, in the much-studied E. coli system is determined by plasmid genes, and is an adaptation for transferring copies of the plasmid from one bacterial host to another. It is seldom that a conjugative plasmid integrates into the host bacterial chromosome, and subsequently transfers part of the host bacterial DNA to another bacterium. Plasmid-mediated transfer of host bacterial DNA also appears to be an accidental process rather than a bacterial adaptation.

Transformation, unlike transduction or conjugation, depends on numerous bacterial gene products that specifically interact to perform this complex process, and thus transformation is clearly a bacterial adaptation for DNA transfer. In order for a bacterium to bind, take up and recombine donor DNA into its own chromosome, it must first enter a special physiological state termed competence. In *Bacillus subtilis*, about 40 genes are required for the development of competence. The length of DNA transferred during *B. subtilis* transformation can be between a third of a chromosome up to the whole chromosome. Transformation appears to be common among bacterial species, and thus far at least 60 species are known to have the nat-ural ability to become competent for transformation. The development of competence in nature is usually associated with stressful environmental conditions, and seems to be an adaptation for facilitating repair of DNA damage in recipient cells.

In ordinary circumstances, transduction, conjugation, and transformation involve transfer of DNA between individual bacteria of the same species, but occasionally transfer may occur between individuals of different bacterial species and this may have significant consequences, such as the transfer of antibiotic resistance. In such cases, gene acquisition from other bacteria or the environment is called *horizontal gene transfer* and may be common under natural conditions. Gene transfer is particularly important in antibiotic resistance as it allows the rapid transfer of resistance genes between different pathogens.

## Bacteriophages

Bacteriophages are viruses that infect bacteria. Many types of bacteriophage exist, some simply infect and lyse their host bacteria, while others insert into the bacterial chromosome. A bacteriophage can contain genes that contribute to its host's phenotype: for example, in the evolution of *Escherichia coli* O157:H7 and *Clostridium botulinum*, the toxin genes in an integrated phage converted a harmless ancestral bacterium into a lethal pathogen. Bacteria resist phage infection through restriction modification systems that degrade foreign DNA, and a system that uses CRISPR sequences to retain fragments of the genomes of phage that the bacteria have come into contact with in the past, which allows them to block virus replication through a form of RNA interference. This CRISPR system provides bacteria with acquired immunity to infection.

## Behavior

## Secretion

Bacteria frequently secrete chemicals into their environment in order to modify it favorably. The secretions are often proteins and may act as enzymes that digest some form of food in the environment.

# Bioluminescence

A few bacteria have chemical systems that generate light. This bioluminescence often occurs in bacteria that live in association with fish, and the light probably serves to attract fish or other large animals.

# Multicellularity

Bacteria often function as multicellular aggregates known as biofilms, exchanging a variety of molecular signals for inter-cell communication, and engaging in coordinated multicellular behavior.

The communal benefits of multicellular cooperation include a cellular division of labor, accessing resources that cannot effectively be utilized by single cells, collectively defending against antagonists, and optimizing population survival by differentiating into distinct cell types. For example, bacteria in biofilms can have more than 500 times increased resistance to antibacterial agents than individual "planktonic" bacteria of the same species.

One type of inter-cellular communication by a molecular signal is called quorum sensing, which serves the purpose of determining whether there is a local population density that is sufficiently high that it is productive to invest in processes that are only successful if large numbers of similar organisms behave similarly, as in excreting digestive enzymes or emitting light.

Quorum sensing allows bacteria to coordinate gene expression, and enables them to produce, release and detect autoinducers or pheromones which accumulate with the growth in cell population.

# Movement

Many bacteria can move using a variety of mechanisms: flagella are used for swimming through fluids; bacterial gliding and twitching motility move bacteria across surfaces; and changes of buoyancy allow vertical motion.

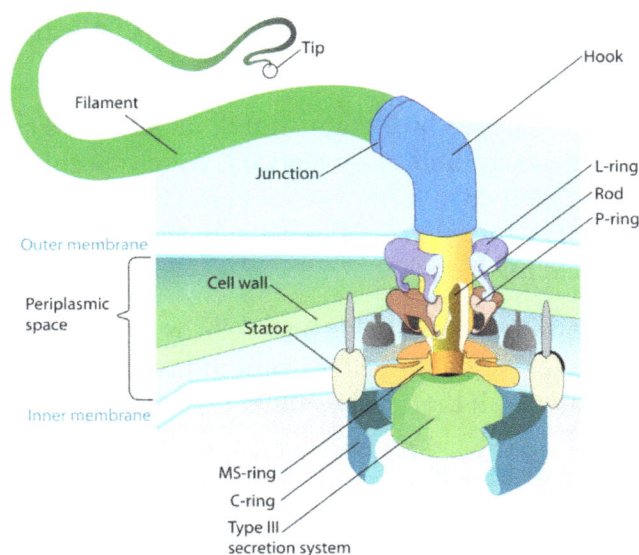

Flagellum of gram-negative bacteria. The base drives the rotation of the hook and filament.

Swimming bacteria frequently move near 10 body lengths per second and a few as fast as 100. This makes them at least as fast as fish, on a relative scale.

In bacterial gliding and twitching motility, bacteria use their *type IV pili* as a grappling hook, repeatedly extending it, anchoring it and then retracting it with remarkable force (>80 pN).

*"Our observations redefine twitching motility as a rapid, highly organized mechanism of bacterial translocation by which Pseudomonas aeruginosa can disperse itself over large areas to colonize new territories. It is also now clear, both morphologically and genetically, that twitching motility and social gliding motility, such as occurs in Myxococcus xanthus, are essentially the same process."*

*— "A re-examination of twitching motility in Pseudomonas aeruginosa" – Semmler, Whitchurch & Mattick (1999)*

*Flagella* are semi-rigid cylindrical structures that are rotated and function much like the propeller on a ship. Objects as small as bacteria operate a low Reynolds number and cylindrical forms are more efficient than the flat, paddle-like, forms appropriate at human-size scale.

Bacterial species differ in the number and arrangement of flagella on their surface; some have a single flagellum (*monotrichous*), a flagellum at each end (*amphitrichous*), clusters of flagella at the poles of the cell (*lophotrichous*), while others have flagella distributed over the entire surface of the cell (*peritrichous*). The bacterial flagella is the best-understood motility structure in any organism and is made of about 20 proteins, with approximately another 30 proteins required for its regulation and assembly. The flagellum is a rotating structure driven by a reversible motor at the base that uses the electrochemical gradient across the membrane for power. This motor drives the motion of the filament, which acts as a propeller.

Many bacteria (such as *E. coli*) have two distinct modes of movement: forward movement (swimming) and tumbling. The tumbling allows them to reorient and makes their movement a three-di-mensional random walk. The flagella of a unique group of bacteria, the spirochaetes, are found between two membranes in the periplasmic space. They have a distinctive helical body that twists about as it moves.

Motile bacteria are attracted or repelled by certain stimuli in behaviors called taxes: these include chemotaxis, phototaxis, energy taxis, and magnetotaxis. In one peculiar group, the myxobacteria, individual bacteria move together to form waves of cells that then differentiate to form fruiting bodies containing spores. The myxobacteria move only when on solid surfaces, unlike *E. coli*, which is motile in liquid or solid media.

Several *Listeria* and *Shigella* species move inside host cells by usurping the cytoskeleton, which is normally used to move organelles inside the cell. By promoting actin polymerization at one pole of their cells, they can form a kind of tail that pushes them through the host cell's cytoplasm.

## Classification and Identification

Classification seeks to describe the diversity of bacterial species by naming and grouping organisms based on similarities. Bacteria can be classified on the basis of cell structure, cellular metabolism or on differences in cell components, such as DNA, fatty acids, pigments, antigens and quinones. While these schemes allowed the identification and classification of bacterial strains, it was unclear whether these differences represented variation between distinct species or between strains of the same species. This uncertainty was due to the lack of distinctive structures in most bacteria, as well as lateral gene transfer between unrelated species. Due to lateral gene transfer, some closely related bacteria can have very different morphologies and metabolisms. To overcome this uncertainty, modern bacterial classification emphasizes molecular systematics, using genetic techniques such as guanine cytosine ratio determination, genome-genome hybridization, as well as sequencing genes that have not undergone extensive lateral gene transfer, such as the rRNA gene. Classification of bacteria is determined by publication in the International Journal of Systematic Bacteriology, and Bergey's Manual of Systematic Bacteriology. The International Committee on Systematic Bacteriology (ICSB) maintains international rules for the naming of bacteria and taxonomic categories and for the ranking of them in the International Code of Nomenclature of Bacteria.

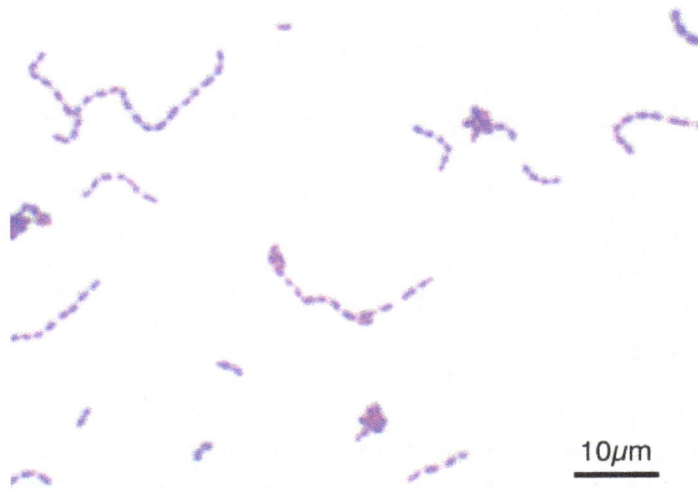

*Streptococcus mutans* visualized with a Gram stain

The term "bacteria" was traditionally applied to all microscopic, single-cell prokaryotes. However, molecular systematics showed prokaryotic life to consist of two separate domains, originally called *Eubacteria* and *Archaebacteria*, but now called *Bacteria* and *Archaea* that evolved independently from an ancient common ancestor. The archaea and eukaryotes are more closely related to each other than either is to the bacteria. These two domains, along with Eukarya, are the basis of the three-domain system, which is currently the most widely used classification system in microbiolology. However, due to the relatively recent introduction of molecular systematics and a rapid increase in the number of genome sequences that are available, bacterial classification remains a changing and expanding field. For example, a few biologists argue that the Archaea and Eukaryotes evolved from gram-positive bacteria.

The identification of bacteria in the laboratory is particularly relevant in medicine, where the correct treatment is determined by the bacterial species causing an infection. Consequently, the need to identify human pathogens was a major impetus for the development of techniques to identify bacteria.

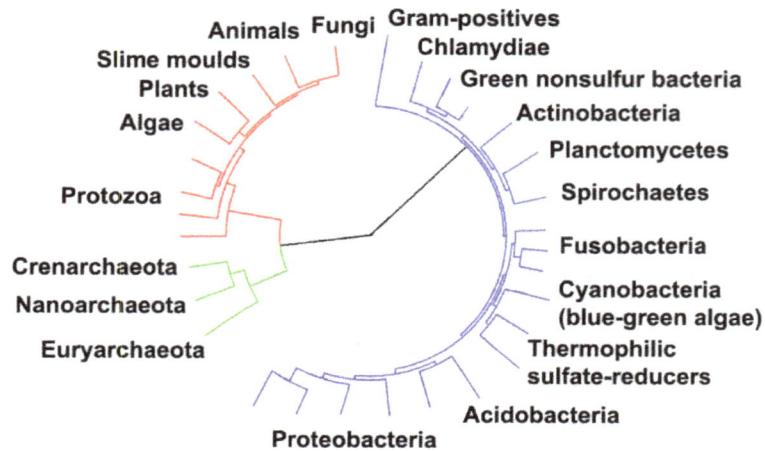

Phylogenetic tree showing the diversity of bacteria, compared to other organisms. Eukaryotes are colored red, archaea green and bacteria blue.

The *Gram stain*, developed in 1884 by Hans Christian Gram, characterises bacteria based on the structural characteristics of their cell walls. The thick layers of peptidoglycan in the "gram-positive" cell wall stain purple, while the thin "gram-negative" cell wall appears pink. By combining morphology and Gram-staining, most bacteria can be classified as belonging to one of four groups (gram-positive cocci, gram-positive bacilli, gram-negative cocci and gram-negative bacilli). Some organisms are best identified by stains other than the Gram stain, particularly mycobacteria or *Nocardia*, which show acid-fastness on Ziehl–Neelsen or similar stains. Other organisms may need to be identified by their growth in special media, or by other techniques, such as serology.

Culture techniques are designed to promote the growth and identify particular bacteria, while restricting the growth of the other bacteria in the sample. Often these techniques are designed for specific specimens; for example, a sputum sample will be treated to identify organisms that cause pneumonia, while stool specimens are cultured on selective media to identify organisms that cause diarrhoea, while preventing growth of non-pathogenic bacteria. Specimens that are normally sterile, such as blood, urine or spinal fluid, are cultured under conditions designed to grow all possible organisms. Once a pathogenic organism has been isolated, it can be further characterised by its morphology, growth patterns (such as aerobic or anaerobic growth), patterns of hemolysis, and staining.

As with bacterial classification, identification of bacteria is increasingly using molecular methods. Diagnostics using DNA-based tools, such as polymerase chain reaction, are increasingly popular due to their specificity and speed, compared to culture-based methods. These methods also allow the detection and identification of "viable but nonculturable" cells that are metabolically active but non-dividing. However, even using these improved methods, the total number of bacterial species is not known and cannot even be estimated with any certainty. Following present classification, there are a little less than 9,300 known species of prokaryotes, which includes bacteria and archaea; but attempts to estimate the true number of bacterial diversity have ranged from $10^7$ to $10^9$ total species – and even these diverse estimates may be off by many orders of magnitude.

## Interactions with Other Organisms

Despite their apparent simplicity, bacteria can form complex associations with other organisms. These symbiotic associations can be divided into parasitism, mutualism and commensalism. Due to their small size, commensal bacteria are ubiquitous and grow on animals and plants exactly as they will grow on any other surface. However, their growth can be increased by warmth and sweat, and large populations of these organisms in humans are the cause of body odor.

## Predators

Some species of bacteria kill and then consume other microorganisms, these species are called *predatory bacteria*. These include organisms such as *Myxococcus xanthus*, which forms swarms of cells that kill and digest any bacteria they encounter. Other bacterial predators either attach to their prey in order to digest them and absorb nutrients, such as *Vampirovibrio chlorellavorus*, or invade another cell and multiply inside the cytosol, such as *Daptobacter*. These predatory bacteria are thought to have evolved from saprophages that consumed dead microorganisms, through adaptations that allowed them to entrap and kill other organisms.

## Mutualists

Certain bacteria form close spatial associations that are essential for their survival. One such mutualistic association, called interspecies hydrogen transfer, occurs between clusters of anaerobic bacteria that consume organic acids, such as butyric acid or propionic acid, and produce hydrogen, and methanogenic Archaea that consume hydrogen. The bacteria in this association are unable to consume the organic acids as this reaction produces hydrogen that accumulates in their surroundings. Only the intimate association with the hydrogen-consuming Archaea keeps the hydrogen concentration low enough to allow the bacteria to grow.

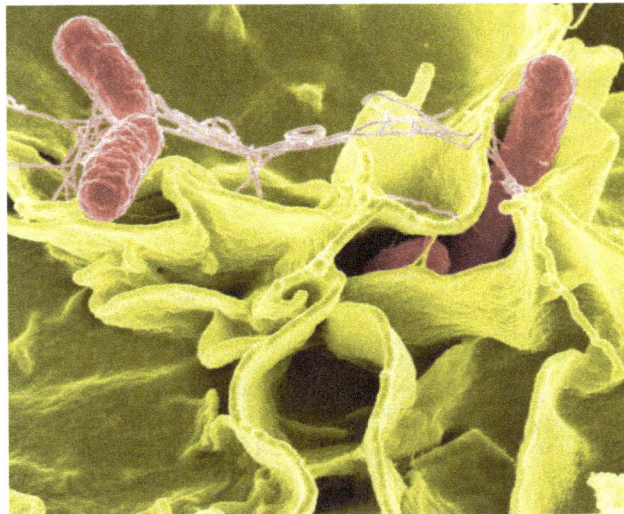

Color-enhanced scanning electron micrograph showing *Salmonella typhimurium* (red) invading cultured human cells

In soil, microorganisms that reside in the rhizosphere (a zone that includes the root surface and the soil that adheres to the root after gentle shaking) carry out nitrogen fixation, converting nitrogen gas to nitrogenous compounds. This serves to provide an easily absorbable form of nitrogen for many plants, which cannot fix nitrogen themselves. Many other bacteria are found as symbionts in humans and other organisms. For example, the presence of over 1,000 bacterial species in the normal human gut flora of the intestines can contribute to gut immunity, synthesise vitamins, such as folic acid, vitamin K and biotin, convert sugars to lactic acid, as well as fermenting complex undigestible carbohydrates. The presence of this gut flora also inhibits the growth of potentially pathogenic bacteria (usually through competitive exclusion) and these beneficial bacteria are consequently sold as probiotic dietary supplements.

## Pathogens

If bacteria form a parasitic association with other organisms, they are classed as pathogens. Pathogenic bacteria are a major cause of human death and disease and cause infections such as tetanus, typhoid fever, diphtheria, syphilis, cholera, foodborne illness, leprosy and tuberculosis. A pathogenic cause for a known medical disease may only be discovered many years after, as was the case with *Helicobacter pylori* and peptic ulcer disease. Bacterial diseases are also important in agriculture, with bacteria causing leaf spot, fire blight and wilts in plants, as well as Johne's disease, mastitis, salmonella and anthrax in farm animals.

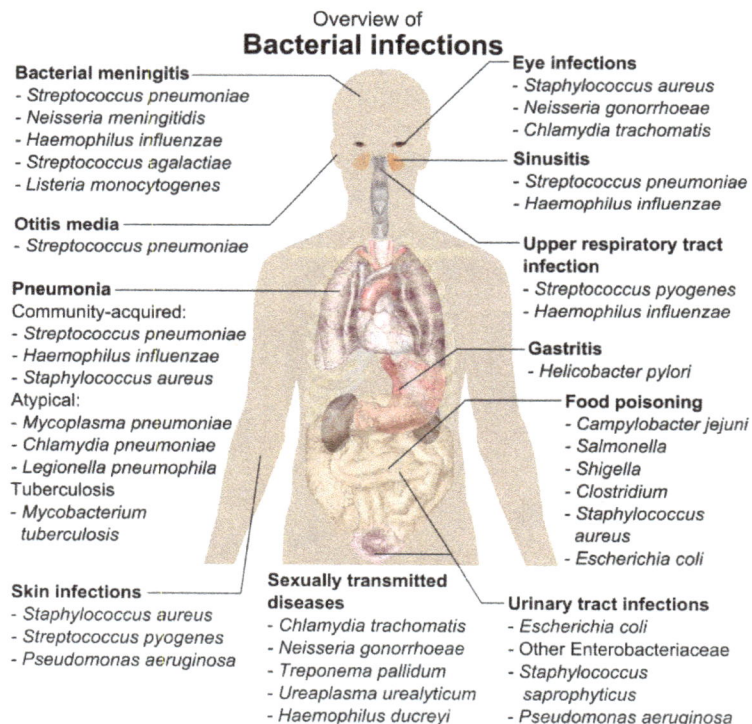

Overview of
### Bacterial infections

**Bacterial meningitis**
- *Streptococcus pneumoniae*
- *Neisseria meningitidis*
- *Haemophilus influenzae*
- *Streptococcus agalactiae*
- *Listeria monocytogenes*

**Otitis media**
- *Streptococcus pneumoniae*

**Pneumonia**
Community-acquired:
- *Streptococcus pneumoniae*
- *Haemophilus influenzae*
- *Staphylococcus aureus*
Atypical:
- *Mycoplasma pneumoniae*
- *Chlamydia pneumoniae*
- *Legionella pneumophila*
Tuberculosis
- *Mycobacterium tuberculosis*

**Skin infections**
- *Staphylococcus aureus*
- *Streptococcus pyogenes*
- *Pseudomonas aeruginosa*

**Sexually transmitted diseases**
- *Chlamydia trachomatis*
- *Neisseria gonorrhoeae*
- *Treponema pallidum*
- *Ureaplasma urealyticum*
- *Haemophilus ducreyi*

**Eye infections**
- *Staphylococcus aureus*
- *Neisseria gonorrhoeae*
- *Chlamydia trachomatis*

**Sinusitis**
- *Streptococcus pneumoniae*
- *Haemophilus influenzae*

**Upper respiratory tract infection**
- *Streptococcus pyogenes*
- *Haemophilus influenzae*

**Gastritis**
- *Helicobacter pylori*

**Food poisoning**
- *Campylobacter jejuni*
- *Salmonella*
- *Shigella*
- *Clostridium*
- *Staphylococcus aureus*
- *Escherichia coli*

**Urinary tract infections**
- *Escherichia coli*
- Other Enterobacteriaceae
- *Staphylococcus saprophyticus*
- *Pseudomonas aeruginosa*

Overview of bacterial infections and main species involved.

Each species of pathogen has a characteristic spectrum of interactions with its human hosts. Some organisms, such as *Staphylococcus* or *Streptococcus*, can cause skin infections, pneumonia, meningitis and even overwhelming sepsis, a systemic inflammatory response producing shock, massive vasodilation and death. Yet these organisms are also part of the normal human flora and usually exist on the skin or in the nose without causing any disease at all.

Other organisms invariably cause disease in humans, such as the Rickettsia, which are obligate intracellular parasites able to grow and reproduce only within the cells of other organisms. One species of Rickettsia causes typhus, while another causes Rocky Mountain spotted fever. *Chlamydia*, another phylum of obligate intracellular parasites, contains species that can cause pneumonia, or urinary tract infection and may be involved in coronary heart disease. Finally, some species, such as *Pseudomonas aeruginosa*, *Burkholderia cenocepacia*, and *Mycobacterium avium*, are opportunistic pathogens and cause disease mainly in people suffering from immunosuppression or cystic fibrosis.

Bacterial infections may be treated with antibiotics, which are classified as bacteriocidal if they kill bacteria, or bacteriostatic if they just prevent bacterial growth. There are many types of antibiotics and each class inhibits a process that is different in the pathogen from that found in the host. An example of how antibiotics produce selective toxicity are chloramphenicol and puromycin, which inhibit the bacterial ribosome, but not the structurally different eukaryotic ribosome. Antibiotics are used both in treating human disease and in intensive farming to promote animal growth, where they may be contributing to the rapid development of antibiotic resistance in bacterial populations. Infections can be prevented by antiseptic measures such as sterilizing the skin prior to piercing it with the needle of a syringe, and by proper care of indwelling catheters. Surgical and dental instruments are also sterilized to prevent contamination by bacteria. Disinfectants such as bleach are used to kill bacteria or other pathogens on surfaces to prevent contamination and further reduce the risk of infection.

## Significance in Technology and Industry

Bacteria, often lactic acid bacteria, such as *Lactobacillus* and *Lactococcus*, in combination with yeasts and molds, have been used for thousands of years in the preparation of fermented foods, such as cheese, pickles, soy sauce, sauerkraut, vinegar, wine and yogurt.

The ability of bacteria to degrade a variety of organic compounds is remarkable and has been used in waste processing and bioremediation. Bacteria capable of digesting the hydrocarbons in petroleum are often used to clean up oil spills. Fertilizer was added to some of the beaches in Prince William Sound in an attempt to promote the growth of these naturally occurring bacteria after the 1989 *Exxon Valdez* oil spill. These efforts were effective on beaches that were not too thickly covered in oil. Bacteria are also used for the bioremediation of industrial toxic wastes. In the chemical industry, bacteria are most important in the production of enantiomerically pure chemicals for use as pharmaceuticals or agrichemicals.

Bacteria can also be used in the place of pesticides in the biological pest control. This commonly involves *Bacillus thuringiensis* (also called BT), a gram-positive, soil dwelling bacterium. Subspecies of this bacteria are used as a Lepidopteran-specific insecticides under trade names such as Dipel and Thuricide. Because of their specificity, these pesticides are regarded as environmentally friendly, with little or no effect on humans, wildlife, pollinators and most other beneficial insects.

Because of their ability to quickly grow and the relative ease with which they can be manipulated, bacteria are the workhorses for the fields of molecular biology, genetics and biochemistry. By making mutations in bacterial DNA and examining the resulting phenotypes, scientists can determine the function of genes, enzymes and metabolic pathways in bacteria, then apply this knowledge to more complex organisms. This aim of understanding the biochemistry of a cell reaches its most complex expression in the synthesis of huge amounts of enzyme kinetic and gene expression data into mathematical models of entire organisms. This is achievable in some well-studied bacteria, with models of *Escherichia coli* metabolism now being produced and tested. This understanding of bacterial metabolism and genetics allows the use of biotechnology to bioengineer bacteria for the production of therapeutic proteins, such as insulin, growth factors, or antibodies.

Because of their importance for research in general, samples of bacterial strains are isolated and preserved in Biological Resource Centers. This ensures the availability of the strain to scientists worldwide.

## History of Bacteriology

Antonie van Leeuwenhoek, the first microbiologist and the first person to observe bacteria using a microscope.

Bacteria were first observed by the Dutch microscopist Antonie van Leeuwenhoek in 1676, using a single-lens microscope of his own design. He then published his observations in a series of letters to the Royal Society of London. Bacteria were Leeuwenhoek's most remarkable microscopic discovery. They were just at the limit of what his simple lenses could make out and, in one of the most striking hiatuses in the history of science, no one else would see them again for over a century. Only then were his by-then-largely-forgotten observations of bacteria — as opposed to his famous "animalcules" (spermatozoa) — taken seriously.

Christian Gottfried Ehrenberg introduced the word "bacterium" in 1828. In fact, his *Bacterium* was a genus that contained non-spore-forming rod-shaped bacteria, as opposed to *Bacillus*, a genus of spore-forming rod-shaped bacteria defined by Ehrenberg in 1835.

Louis Pasteur demonstrated in 1859 that the growth of microorganisms causes the fermentation process, and that this growth is not due to spontaneous generation. (Yeasts and molds, commonly associated with fermentation, are not bacteria, but rather fungi.) Along with his contemporary Robert Koch, Pasteur was an early advocate of the germ theory of disease.

Robert Koch, a pioneer in medical microbiology, worked on cholera, anthrax and tuberculosis. In his research into tuberculosis Koch finally proved the germ theory, for which he received a Nobel Prize in 1905. In *Koch's postulates*, he set out criteria to test if an organism is the cause of a disease, and these postulates are still used today.

Though it was known in the nineteenth century that bacteria are the cause of many diseases, no effective antibacterial treatments were available. In 1910, Paul Ehrlich developed the first antibiotic, by changing dyes that selectively stained *Treponema pallidum* — the spirochaete that causes syphilis — into compounds that selectively killed the pathogen. Ehrlich had been awarded a 1908 Nobel Prize for his work on immunology, and pioneered the use of stains to detect and identify bacteria, with his work being the basis of the Gram stain and the Ziehl–Neelsen stain.

A major step forward in the study of bacteria came in 1977 when Carl Woese recognized that archaea have a separate line of evolutionary descent from bacteria. This new phylogenetic taxonomy depended on the sequencing of 16S ribosomal RNA, and divided prokaryotes into two evolutionary domains, as part of the three-domain system.

## Bacterial Plant Pathogens

### Burkholderia

*Burkholderia* is a genus of Proteobacteria whose pathogenic members include *Burkholderia mallei*, responsible for glanders, a disease that occurs mostly in horses and related animals; *Burkholderia pseudomallei*, causative agent of melioidosis; and *Burkholderia cepacia*, an important pathogen of pulmonary infections in people with cystic fibrosis (CF).

The *Burkholderia* (previously part of *Pseudomonas*) genus name refers to a group of virtually ubiquitous Gram-negative, motile, obligately aerobic, rod-shaped bacteria including both animal and plant pathogens, as well as some environmentally important species. In particular, *B. xenovorans* (previously named *Pseudomonas cepacia* then *B. cepacia* and *B. fungorum*) is renowned for being catalase positive (affecting patients with chronic granulomatous disease) and its ability to degrade chlororganic pesticides and polychlorinated biphenyls (PCBs).

The use of *Burkholderia* species for agricultural purposes (such as biodegradation, biocontrol, and plant growth-promoting rhizobacteria) is subject to discussions because of possible pathogenic effects in immunocompromised people (especially CF sufferers), e.g., hospital-acquired infections. However, the animal pathogenic and the plant/soil species belong to different groups, and it was proposed to separate them into two different genera, to avoid misinterpretations.

Due to their antibiotic resistance and the high mortality rate from their associated diseases, *B. mallei* and *B. pseudomallei* are considered to be potential biological warfare agents, targeting livestock and humans.

The genus was named after Walter H. Burkholder, plant pathologist at Cornell University.

### Xanthomonas

*Xanthomonas* is a genus of Proteobacteria, many of which cause plant diseases.

## Taxonomy

The *Xanthomonas* genus has been subject of numerous taxonomic and phylogenetic studies and was first described as *Bacterium vesicatorium* as a pathogen of pepper and tomato in 1921. Dowson later reclassified the bacterium as *Xanthomonas campestris* and proposed the genus *Xanthomonas.Xanthomonas* was first described as a monotypic genus and further research resulted in the division into two groups, A and B. Later work using DNA:DNA hybridization has served as a framework for the general *Xanthomonas* species classification. Other tools, including multilocus sequence analysis and amplified fragment-length polymorphism, have been used for classification within clades. While previous research has illustrated the complexity of the *Xanthomonas* genus, recent research appears to have resulted in a clearer picture. More recently, genome-wide analysis of multiple *Xanthomonas* strains mostly supports the previous phylogenies.

## Morphology and Growth

Individual cell characteristics include:

- Cell type – straight rods

- Size – 0.4 – 1.0 µm wide by 1.2 – 3.0 µm long

- Motility – motile by a single polar flagellum

Colony growth characteristics include:

- Mucoid, convex, and yellow colonies on YDC medium

- Yellow pigment from xanthomonadin, which contains bromine

- Most produce large amounts of extracellular polysaccharide

- Temperature range – 4 to 37 °C

Biochemical and physiological test results are:

- Gram stain – negative

- Catalase positive

- Oxidase negative

## Xanthomonas Plant Pathogens

*Xanthomonas* species can cause bacterial spots and blights of leaves, stems, and fruits on a wide variety of plant species. Pathogenic species show high degrees of specificity and some are split into multiple pathovars, a species designation based on host specificity.

Bacterial blight of cotton, caused by *Xanthomonas citri* subsp. *malvacearum* is the most important bacterial disease on cotton which infects all aerial parts of the host. Loss due to this disease was estimated for about 10 to 30% on different cultivars and can be found in Asia, Africa and southern America. Citrus canker, caused by *Xanthomonas citri* subsp. *citri* is an economically important disease of many citrus species (lime, orange, lemon, pamelo, etc.)

Bacterial leaf spot has caused significant crop losses over the years. Causes of this disease include *Xanthomonas euvesicatoria* and *Xanthomonas perforans* = [*Xanthomonas axonopodis* (syn. *campestris*) pv. *vesicatoria*], *Xanthomonas vesicatoria*, and *Xanthomonas gardneri*. In some areas where infection begins soon after transplanting, the total crop can be lost as a result of this disease.

Bacterial blight of rice, caused by *Xanthomonas oryzae* pv. *oryzae*, is a disease found worldwide and particularly destructive in the rice-producing regions in Asia.

## Plant Pathogenesis and Disease Control

*Xanthomonas* species can be easily spread in water, movement of infected material such as seed or propagation plants, and by mechanical means such as infected pruning tools. Upon contact with a susceptible host, bacteria enter through wounds or natural plant openings as a means to infect. They inject a number of effector proteins, including TAL effectors, into the plant by their secretion systems (i.e., type III secretion system).

To prevent infections, limiting the introduction of the bacteria is key. Some resistant cultivars of certain plant species are available as this may be the most economical means for controlling this disease. For chemical control, preventative applications are best to reduce the potential for bacterial development. Copper-containing products offer some protection along with field-grade antibiotics such as oxytetracycline, which is labeled for use on some food crops in the United States. Curative applications of chemical pesticides may slow or reduce the spread of the bacterium, but will not cure already diseased plants. It is important to consult chemical pesticide labels when attempting to control bacterial diseases, as different *Xanthomonas* species can have different responses to these applications. Over-reliance on chemical control methods can also result in the selection of resistant isolates, so these applications should be considered a last resort.

## Industrial Use

*Xanthomonas* species produce an extrapolysaccharide called xanthan gum that has a wide range of industrial uses, including foods, petroleum products, and cosmetics.

## *Xanthomonas* Resources

Isolates of most species of *Xanthomonas* are available from the National Collection of Plant Pathogenic Bacteria in the United Kingdom and other international culture collections such as ICMP in New Zealand, CFBP in France, and VKM in Russia. It also can be taken out from MTCC India.

Multiple genomes of *Xanthomonas* have been sequenced and additional data sets/tools are available at The *Xanthomonas* Resource.

## Pseudomonas Tomato

'*Pseudomonas tomato*' is a Gram-negative plant pathogenic bacterium that infects a variety of plants. It was once considered a pathovar of *Pseudomonas syringae*, but following DNA-relatedness studies, it was recognized as a separate species and several other former *P. syringae* pathovars were incorporated into it. Since no official name has yet been given, it is referred to by the epithet '*Pseudomonas tomato*'.

# Plant Virus

Pepper mild mottle virus

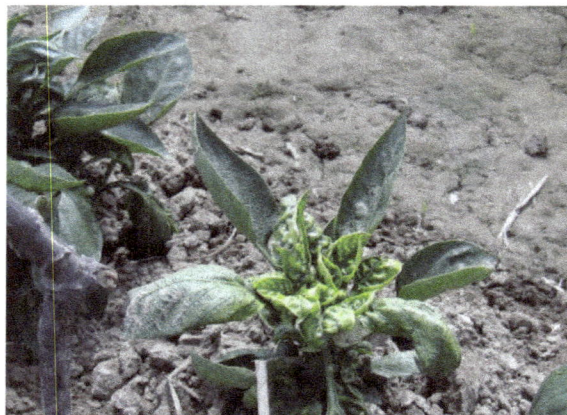

Leaf curl virus

Plant viruses are viruses that affect plants. Like all other viruses, plant viruses are obligate intracellular parasites that do not have the molecular machinery to replicate without a host. Plant
viruses are pathogenic to higher plants. While this article does not intend to list all plant viruses, it
discusses some important viruses as well as their uses in plant molecular biology.

## Overview

Although plant viruses are not nearly as well understood as the animal counterparts, one plant
virus has become iconic. The first virus to be discovered was *Tobacco mosaic virus* (TMV). This
and other viruses cause an estimated US$60 billion loss in crop yields worldwide each year. Plant
viruses are grouped into 73 genera and 49 families. However, these figures relate only to
cultivated plants that represent only a tiny fraction of the total number of plant species. Viruses
in wild plants have been poorly studied, but those studies that exist almost overwhelming show
that such interactions between wild plants and their viruses do not appear to cause disease in the
host plants.

To transmit from one plant to another and from one plant cell to another, plant viruses must use strategies that are usually different from animal viruses. Plants do not move, and so plant-to-plant transmission usually involves vectors (such as insects). Plant cells are surrounded by solid cell walls, therefore transport through plasmodesmata is the preferred path for virions to move between plant cells. Plants probably have specialized mechanisms for transporting mRNAs through plasmodesmata, and these mechanisms are thought to be used by RNA viruses to spread from one cell to another.

Plant defenses against viral infection include, among other measures, the use of siRNA in response to dsRNA. Most plant viruses encode a protein to suppress this response. Plants also reduce transport through plasmodesmata in response to injury.

## History

The discovery of plant viruses causing disease is often accredited to A. Mayer (1886) working in the Netherlands demonstrated that the sap of mosaic obtained from tobacco leaves developed mosaic symptom when injected in healthy plants. However the infection of the sap was destroyed when it was boiled. He thought that the causal agent was the bacteria. However, after larger inoculation with a large number of bacteria, he failed to develop a mosaic symptom.

In 1898, Martinus Beijerinck, who was a Professor of Microbiology at the Technical University the Netherlands, put forth his concepts that viruses were small and determined that the "mosaic disease" remained infectious when passed through a Chamberland filter-candle. This was in contrast to bacteria microorganisms, which were retained by the filter. Beijerinck referred to the infectious filtrate as a "contagium vivum fluidum", thus the coinage of the modern term "virus".

After the initial discovery of the 'viral concept' there was need to classify any other known viral diseases based on the mode of transmission even though microscopic observation proved fruitless. In 1939 Holmes published a classification list of 129 plant viruses. This was expanded and in 1999 there were 977 officially recognized, and some provisional, plant virus species.

The purification (crystallization) of TMV was first performed by Wendell Stanley, who published his findings in 1935, although he did not determine that the RNA was the infectious material. However, he received the Nobel Prize in Chemistry in 1946. In the 1950s a discovery by two labs simultaneously proved that the purified RNA of the TMV was infectious which reinforced the argument. The RNA carries genetic information to code for the production of new infectious particles.

More recently virus research has been focused on understanding the genetics and molecular biology of plant virus genomes, with a particular interest in determining how the virus can replicate, move and infect plants. Understanding the virus genetics and protein functions has been used to explore the potential for commercial use by biotechnology companies. In particular, viral-derived sequences have been used to provide an understanding of novel forms of resistance. The recent boom in technology allowing humans to manipulate plant viruses may provide new strategies for production of value-added proteins in plants.

## Structure

Viruses are extremely small and can only be observed under an electron microscope. The structure of a virus is given by its coat of proteins, which surround the viral genome. Assembly of viral particles takes place spontaneously.

Over 50% of known plant viruses are rod-shaped (flexuous or rigid). The length of the particle is normally dependent on the genome but it is usually between 300–500 nm with a diameter of 15–20 nm. Protein subunits can be placed around the circumference of a circle to form a disc. In the presence of the viral genome, the discs are stacked, then a tube is created with room for the nucleic acid genome in the middle.

The second most common structure amongst plant viruses are isometric particles. They are 25–50 nm in diameter. In cases when there is only a single coat protein, the basic structure consists of 60 T subunits, where T is an integer. Some viruses may have 2 coat proteins that associate to form an icosahedral shaped particle.

There are three genera of *Geminiviridae* that possess geminate particles which are like two isometric particles stuck together.

A very small number of plant viruses have, in addition to their coat proteins, a lipid envelope. This is derived from the plant cell membrane as the virus particle buds off from the cell.

## Transmission of Plant Viruses

## Through Sap

Viruses can be spread by direct transfer of sap by contact of a wounded plant with a healthy one. Such contact may occur during agricultural practices, as by damage caused by tools or hands, or naturally, as by an animal feeding on the plant. Generally TMV, potato viruses and cucumber mosaic viruses are transmitted via sap.

## Insects

Plant viruses need to be transmitted by a vector, most often insects such as leafhoppers. One class of viruses, the Rhabdoviridae, has been proposed to actually be insect viruses that have evolved to replicate in plants. The chosen insect vector of a plant virus will often be the determining factor in that virus's host range: it can only infect plants that the insect vector feeds upon. This was shown in part when the old world white fly made it to the USA, where it transferred many plant viruses into new hosts. Depending on the way they are transmitted, plant viruses are classified as non-persistent, semi-persistent and persistent. In non-persistent transmission, viruses become attached to the distal tip of the stylet of the insect and on the next plant it feeds on, it inoculates it with the virus. Semi-persistent viral transmission involves the virus entering the foregut of the insect. Those viruses that manage to pass through the gut into the haemolymph and then to the salivary glands are known as persistent. There are two sub-classes of persistent viruses: propagative and circulative. Propagative viruses are able to replicate in both the plant and the insect (and may have originally been insect viruses), whereas circulative can not. Circulative viruses are protected inside aphids by the chaperone protein symbionin, produced by bacterial symbionts. Many plant viruses encode within their genome polypeptides with domains essential for transmission by insects. In non-persistent and semi-persistent viruses, these domains are in the coat protein and another protein known as the helper component. A bridging hypothesis has been proposed to explain how these proteins aid in insect-mediated viral transmission. The helper component will bind to the specific domain of the coat protein, and then the insect mouthparts — creating a bridge. In persistent propagative viruses, such as tomato spotted wilt virus (TSWV), there is often a lipid coat surrounding the proteins that is not seen in other classes of plant viruses. In the case of TSWV, 2 viral proteins are expressed in this lipid envelope. It has been proposed that the viruses bind via these proteins and are then taken into the insect cell by receptor-mediated endocytosis.

## Nematodes

Soil-borne nematodes also have been shown to transmit viruses. They acquire and transmit them by feeding on infected roots. Viruses can be transmitted both non-persistently and persistently, but there is no evidence of viruses being able to replicate in nematodes. The virions attach to the stylet (feeding organ) or to the gut when they feed on an infected plant and can then detach during later feeding to infect other plants. Examples of viruses that can be transmitted by nematodes include tobacco ringspot virus and tobacco rattle virus.

## Plasmodiophorids

A number of virus genera are transmitted, both persistently and non-persistently, by soil borne zoosporic protozoa. These protozoa are not phytopathogenic themselves, but parasitic. Transmission of the virus takes place when they become associated with the plant roots. Examples include *Polymyxa graminis*, which has been shown to transmit plant viral diseases in cereal crops and *Polymyxa betae* which transmits Beet necrotic yellow vein virus. Plasmodiophorids also create wounds in the plant's root through which other viruses can enter.

## Seed and Pollen Borne Viruses

Plant virus transmission from generation to generation occurs in about 20% of plant viruses. When viruses are transmitted by seeds, the seed is infected in the generative cells and the virus is maintained in the germ cells and sometimes, but less often, in the seed coat. When the growth and development of plants is delayed because of situations like unfavourable weather, there is an increase in the amount of virus infections in seeds. There does not seem to be a correlation between the location of the seed on the plant and its chances of being infected. Little is known about the mechanisms involved in the transmission of plant viruses via seeds, although it is known that it is environmentally influenced and that seed transmission occurs because of a direct invasion of the embryo via the ovule or by an indirect route with an attack on the embryo mediated by infected gametes. These processes can occur concurrently or separately depending on the host plant. It is unknown how the virus is able to directly invade and cross the embryo and boundary between the parental and progeny generations in the ovule. Many plants species can be infected through seeds including but not limited to the families Leguminosae, Solanaceae, Compositae, Rosaceae, Cucurbitaceae, Gramineae. Bean common mosaic virus is transmitted through seeds.

## Direct Plant-to-human Transmission

Researchers from the University of the Mediterranean in Marseille, France have found tenuous evidence that suggest a virus common to peppers, the Pepper Mild Mottle Virus (PMMoV) may have moved on to infect humans. This is a very rare and highly unlikely event as, to enter a cell and replicate, a virus must "bind to a receptor on its surface, and a plant virus would be highly unlikely to recognize a receptor on a human cell. One possibility is that the virus does not infect human cells directly. Instead, the naked viral RNA may alter the function of the cells through a mechanism similar to RNA interference, in which the presence of certain RNA sequences can turn genes on and off," according to Virologist Robert Garry from the Tulane University in New Orleans, Louisiana.

## Translation of Plant Viral Proteins

75% of plant viruses have genomes that consist of single stranded RNA (ssRNA). 65% of plant viruses have +ssRNA, meaning that they are in the same sense orientation as messenger RNA but 10% have -ssRNA, meaning they must be converted to +ssRNA before they can be translated. 5% are double stranded RNA and so can be immediately translated as +ssRNA viruses. 3% require a reverse transcriptase enzyme to convert between RNA and DNA. 17% of plant viruses are ssDNA and very few are dsDNA, in contrast a quarter of animal viruses are dsDNA and three quarters of bacteriophage are dsDNA. Viruses use the plant ribosomes to produce the 4-10 proteins encoded by their genome. However, since many of the proteins are encoded on a single strand (that is, they are polycistronic) this will mean that the ribosome will either only produce one protein, as it will terminate translation at the first stop codon, or that a polyprotein will be produced. Plant viruses have had to evolve special techniques to allow the production of viral proteins by plant cells.

## 5' Cap

For translation to occur, eukaryotic mRNAs require a 5' Cap structure. This means that viruses must also have one. This normally consists of 7MeGpppN where N is normally adenine or guanine. The viruses encode a protein, normally a replicase, with a methyltransferase activity to allow this.

Some viruses are cap-snatchers. During this process, a $^{7m}$G-capped host mRNA is recruited by the viral transcriptase complex and subsequently cleaved by a virally encoded endonuclease. The resulting capped leader RNA is used to prime transcription on the viral genome.

However some plant viruses do not use cap, yet translate efficiently due to cap-independent translation enhancers present in 5' and 3' untranslated regions of viral mRNA.

## Readthrough

Some viruses (e.g. tobacco mosaic virus (TMV)) have RNA sequences that contain a "leaky" stop codon. In TMV 95% of the time the host ribosome will terminate the synthesis of the polypeptide at this codon but the rest of the time it continues past it. This means that 5% of the proteins produced are larger than and different from the others normally produced, which is a form of translational regulation. In TMV, this extra sequence of polypeptide is an RNA polymerase that replicates its genome.

## Production of Sub-genomic RNAs

Some viruses use the production of subgenomic RNAs to ensure the translation of all proteins within their genomes. In this process the first protein encoded on the genome, and this the first to be translated, is a replicase. This protein will act on the rest of the genome producing negative strand sub-genomic RNAs then act upon these to form positive strand sub-genomic RNAs that are essentially mRNAs ready for translation.

## Segmented Genomes

Some viral families, such as the *Bromoviridae* instead opt to have multipartite genomes, genomes split between multiple viral particles. For infection to occur, the plant must be infected with all particles across the genome. For instance *Brome mosaic virus* has a genome split between 3 viral particles, and all 3 particles with the different RNAs are required for infection to take place.

## Polyprotein Processing

This strategy is adopted by viral genera such as the Potyviridae and Tymoviridae. The ribosome translates a single protein from the viral genome. Within the polyprotein is an enzyme (or enzymes) with proteinase function that is able to cleave the polyprotein into the various single proteins or just cleave away the protease, which can then cleave other polypeptides producing the mature proteins.

## Well Understood Plant Viruses

Tobacco mosaic virus (TMV) and Cauliflower mosaic virus (CaMV) are frequently used in plant molecular biology. Of special interest is the CaMV 35S promoter, which is a very strong promoter most frequently used in plant transformations.

# Viroid

Viroids are among the smallest infectious pathogens known, larger only than prions, which are misfolded proteins. Viroids consist solely of short strands of circular, single-stranded RNA without protein coats. They are mostly plant pathogens, some of which are of economic importance. Viroid genomes are extremely small in size, ranging from 246 to 467 nucleobases. In comparison, the genome of the smallest known viruses capable of causing an infection by themselves are around 2,000 nucleobases in size. The human pathogen hepatitis D virus is a defective RNA virus similar to viroids.

Viroids, the first known representatives of a new domain of "sub-viral pathogens", were discovered, initially characterized, and named by Theodor Otto Diener, plant pathologist at the U.S Department of Agriculture's Research Center in Beltsville, Maryland, in 1971. The first viroid to be identified was *Potato spindle tuber viroid* (PSTVd). Some 33 species have been identified.

Viroids do not code for any protein. Viroid's replication mechanism uses RNA polymerase II, a host cell enzyme normally associated with synthesis of messenger RNA from DNA, which instead catalyzes "rolling circle" synthesis of new RNA using the viroid's RNA as a template. Some viroids are ribozymes, having catalytic properties which allow self-cleavage and ligation of unit-size genomes from larger replication intermediates.

With Diener's 1989 hypothesis that viroids may represent "living relics" from the widely assumed, ancient, and non-cellular RNA world—extant before the evolution of DNA or proteins—viroids have assumed significance beyond plant pathology to evolutionary science, by representing the most plausible RNAs capable of performing crucial steps in abiogenesis, the evolution of life from inanimate matter.

## Taxonomy

- Family Pospiviroidae

    o Genus *Pospiviroid*; type species: *Potato spindle tuber viroid* ; 356–361 nucleotides(nt)

    o Genus *Pospiviroid*; type species: *Citrus exocortis* ; 368–467 nt

    o Genus *Hostuviroid*; type species: *Hop stunt viroid* ; 294–303 nt

    o Genus *Cocadviroid*; type species: *Coconut cadang-cadang viroid*; 246–247 nt

    o Genus *Apscaviroid*; type species: *Apple scar skin viroid* ; 329–334 nt

    o Genus *Coleviroid*; type species: *Coleus blumei viroid 1* ; 248–251 nt

Putative secondary structure of the PSTVd viroid

- Family Avsunviroidae

    o Genus *Avsunviroid*; type species: *Avocado sunblotch viroid* ; 246–251 nt

    o Genus *Pelamoviroid*; type species: *Peach latent mosaic viroid* ;335–351 nt

    o Genus *Elaviroid*; type species: *Eggplant latent viroid* ; 332–335 nt

## Transmissio

The reproduction mechanism of a typical viroid. Leaf contact transmits the viroid. The viroid enters the cell via its plasmodesmata. RNA polymerase II catalyzes rolling-circle synthesis of new viroids.

Viroid infections are transmitted by cross contamination following mechanical damage to plants as a result of horticultural or agricultural practices. Some are transmitted by aphids and they can also be transferred from plant to plant by leaf contact.

## Replication

Viroids replicate in the nucleus (*Pospiviroidae*) or chloroplasts (*Avsunviroidae*) of plant cells in three steps through an RNA-based mechanism. They require RNA polymerase II, a host cell enzyme normally associated with synthesis of messenger RNA from DNA, which instead catalyzes "rolling circle" synthesis of new RNA using the viroid as template Some viroids are ribozymes, having catalytic properties which allow self-cleavage and ligation of unit-size genomes from larger replication intermediates.

## RNA Silencing

There has long been uncertainty over how viroids induce symptoms in plants without encoding any protein products within their sequences. Evidence suggests that RNA silencing is involved in the process. First, changes to the viroid genome can dramatically alter its virulence. This reflects the fact that any siRNAs produced would have less complementary base pairing with target messenger RNA. Secondly, siRNAs corresponding to sequences from viroid genomes have been isolated from infected plants. Finally, transgenic expression of the noninfectious hpRNA of potato spindle tuber viroid develops all the corresponding viroid-like symptoms. This indicates that when viroids replicate via a double stranded intermediate RNA, they are targeted by a dicer enzyme and cleaved into siRNAs that are then loaded onto the RNA-induced silencing complex. The viroid siRNAs contain sequences capable of complementary base pairing with the plant's own messenger RNAs, and induction of degradation or inhibition of translation causes the classic viroid symptoms.

## Living Relics of The RNA World

Diener's 1989 hypothesis proposed that unique properties of viroids make them more plausible macromolecules than introns, or other RNAs considered in the past as possible "living relics" of a hypothetical, pre-cellular RNA world. If so, viroids have assumed significance beyond plant virology for evolutionary science, because their properties make them more plausible candidates than other RNAs to perform crucial steps in the evolution of life from inanimate matter (abiogenesis). These properties are:

1. viroids' small size, imposed by error-prone replication

2. their high guanine and cytosine content, which increases stability and replication fidelity

3. their circular structure, which assures complete replication without genomic tags

4. existence of structural periodicity, which permits modular assembly into enlarged genomes

5. their lack of protein-coding ability, consistent with a ribosome-free habitat

6. replication mediated in some by ribozymes—the fingerprint of the RNA world

Diener's hypothesis was mostly forgotten until 2014, when it was resurrected in a review article by Flores et al., in which the authors summarized Diener's evidence supporting his hypothesis. In the same year, *New York Times* science writer Carl Zimmer published a popularized piece that mistakenly credited Flores et al. with the hypothesis' original conception.

The presence, in extant cells, of RNAs with molecular properties predicted for RNAs of the RNA World constitutes another powerful argument supporting the RNA World hypothesis.

## History

In the 1920s, symptoms of a previously unknown potato disease were noticed in New York and New Jersey fields. Because tubers on affected plants become elongated and misshaped, they named it the potato spindle tuber disease.

The symptoms appeared on plants onto which pieces from affected plants had been budded—indicating that the disease was caused by a transmissible pathogenic agent. However, a fungus or bacterium could not be found consistently associated with symptom-bearing plants, and therefore, it was assumed the disease was caused by a virus. Despite numerous attempts over the years to isolate and purify the assumed virus, using increasingly sophisticated methods, these were unsuccessful when applied to extracts from potato spindle tuber disease-afflicted plants.

In 1971 Theodor O. Diener showed that the agent was not a virus, but a totally unexpected novel type of pathogen, one-80th the size of typical viruses, for which he proposed the term "viroid". Parallel to agriculture-directed studies, more basic scientific research elucidated many of viroids' physical, chemical, and macromolecular properties. Viroids were shown to consist of short stretches (a few hundred nucleobases) of single-stranded RNA and, unlike viruses, did not have a protein coat. Compared with other infectious plant pathogens, viroids are extremely small in size, ranging from 246 to 467 nucleobases; they thus consist of fewer than 10,000 atoms. In comparison, the genomes of the smallest known viruses capable of causing an infection by themselves are around 2,000 nucleobases long.

In 1976, Sänger et al. presented evidence that potato spindle tuber viroid is a "single-stranded, covalently closed, circular RNA molecule, existing as a highly base-paired rod-like structure"—believed to be the first such molecule described. Circular RNA, unlike linear RNA, forms a covalently closed continuous loop, in which the 3' and 5' ends present in linear RNA molecules have been joined together. Sänger et al. also provided evidence for the true circularity of viroids by finding that the RNA could not be phosphorylated at the 5' terminus. Then, in other tests, they failed to find even one free 3' end, which ruled out the possibility of the molecule having two 3' ends. Viroids thus are true circular RNAs.

The single-strandedness and circularity of viroids was confirmed by electron microscopy, and Gross et al. determined the complete nucleotide sequence of potato spindle tuber viroid in 1978. PSTV was the first pathogen of a eukaryotic organism for which the complete molecular structure has been established. Over thirty plant diseases have since been identified as viroid-, not virus-caused, as had been assumed.

# Phytoplasma

Phytoplasmas are specialised bacteria that are obligate parasites of plant phloem tissue and transmitting insects (vectors). They were discovered by scientists in 1967 and were named mycoplasma-like organisms or MLOs. Since their discovery, phytoplasmas have resisted all attempts to be cultured in vitro in any cell free media, hence routine cultivation in artificial media is still to be established. Nevertheless, still under trial stage, phytoplasma growth in specific artificial media has recently been shown. They are characterised by their lack of a cell wall, a pleiomorphic or filamentous shape, normally with a diameter less than 1 μm, and their very small genomes.

Phytoplasmas are pathogens of agriculturally important plants, including coconut, sugarcane, and sandalwood, causing a wide variety of symptoms that range from mild yellowing to death of infected plants. They are most prevalent in tropical and subtropical regions of the world. They require a vector to be transmitted from plant to plant, and this normally takes the form of sap-sucking insects such as leaf hoppers, in which they are also able to survive and replicate.

## History

References to diseases now known to be caused by phytoplasmas occurred as far back as 1603 for mulberry dwarf disease in Japan. Such diseases were originally thought to be caused by viruses, which, like phytoplasmas, require insect vectors, cannot be cultured, and have some symptom similarity. In 1967, phytoplasmas were discovered in ultrathin sections of plant phloem tissue and named mycoplasma-like organisms (MLOs), because they physically resembled mycoplasmas The organisms were renamed phytoplasmas in 1994, at the 10th Congress of The International Organization for Mycoplasmology.

## Morphology

Being Mollicutes, a phytoplasma lacks a cell wall and instead is bound by a triple-layered membrane. The cell membranes of all phytoplasmas studied so far usually contain a single immunodominant protein (of unknown function) that makes up the majority of the protein content of the cell membrane. The typical phytoplasma exhibits a pleiomorphic or filamentous shape and is less than 1 μm in diameter. As prokaryotes, phytoplasmas' DNA is found throughout the cytoplasm, rather than being concentrated in a nucleus.

## Symptoms

A common symptom caused by phytoplasma infection is phyllody, the production of leaf-like structures in place of flowers. Evidence suggests the phytoplasma downregulates a gene involved in petal formation (*AP3* and its orthologues) and genes involved in the maintenance of the apical meristem (*Wus* and *CLV1*). Other symptoms, such as the yellowing of leaves, are thought to be caused by the phytoplasma's presence in the phloem, affecting its function and changing the transport of carbohydrates.

Phytoplasma-infected plants may also suffer from virescence, the development of green flowers due to the loss of pigment in the petal cells. Phytoplasma-harboring plants which are able to flower may nevertheless be sterile. A phytoplasma effector protein (SAP54) has been identified as inducing symptoms of virescence and phyllody when expressed in plants.

Many plants infected by phytoplasmas gain a bushy or "witches' broom" appearance due to changes in their normal growth patterns. Most plants show apical dominance, but phytoplasma infection can cause the proliferation of auxiliary (side) shoots and an increase in size of the internodes. Such symptoms are actually useful in the commercial production of poinsettias. The infection produces more axillary shoots, which enables production of poinsettia plants that have more than one flower.

Phytoplasmas may cause many other symptoms that are induced because of the stress placed on the plant by infection rather than specific pathogenicity of the phytoplasma. Photosynthesis, especially photosystem II, is inhibited in many phytoplasma-infected plants. Phytoplasma-infected plants often show yellowing which is caused by the breakdown of chlorophyll, the biosynthesis of which is also inhibited.

## Effector (Virulence) Proteins

Many plant pathogens produce virulence factors (or effectors) that modulate or interfere with normal host processes in a way that is beneficial to the pathogen. TCP transcription factors normally regulate plant development and control the expression of lipoxygenase (*LOX*) genes that are required for the biosynthesis of jasmonate. In infected *Arabidopsis* plants (and plants that express SAP11 transgenically), jasmonate levels are decreased. The downregulation of jasmonate production is beneficial to the phytoplasma because jasmonate is involved in plant defence against herbivorous insects such as leafhoppers, and leafhoppers have been shown to lay more eggs on AY-WB-infected plants at least in part because of SAP11. For example, the leafhopper *Macrosteles quadrilineatus* lays 30% more eggs on plants that express SAP11 transgenically, and 60% more eggs on plants infected with AY-WB. Phytoplasmas cannot survive in the external environment and are dependent upon insects such as leafhoppers for transmission to new (healthy) plants. Thus, by interfering with jasmonate production, SAP11 'encourages' leafhoppers to lay more eggs on phytoplasma-infected plants, thereby ensuring that newly hatching leafhopper nymphs feed upon infected plants and become vectors for the bacteria.

## Transmission

## Movement between Plants

Phytoplasmas are mainly spread by insects of the families Cicadellidea (leafhoppers), Fulgoridea (planthoppers), and Psyllidae (jumping plant lice) , which feed on the phloem tissues of infected plants, picking up the phytoplasmas and transmitting them to the next plant on which they feed. So, the host range of phytoplasmas is strongly dependent upon its insect vector. Phytoplasmas contain a major antigenic protein that makes up the majority of their cell surface proteins. This protein has been shown to interact with insect microfilament complexes and is believed to be the determining factor in insect-phytoplasma interaction. Phytoplasmas may overwinter in insect vectors or perennial plants. Phytoplasmas can have varying effects on their insect hosts; examples of both reduced and increased fitness have been seen.

Phytoplasmas enter the insect's body through the stylet, move through the intestine, and are then absorbed into the haemolymph. From there they proceed to colonise the salivary glands, a process that can take up to three weeks. Once established, phytoplasmas are found in most major organs of an infected insect host. The time between being taken up by the insect and reaching an infectious titre in the salivary glands is called the latency period.

Phytoplasmas can also be spread via dodders (Cuscuta) or vegetative propagation such as the grafting of a piece of infected plant onto a healthy plant.

## Movement Within Plants

Phytoplasmas are able to move within the phloem from source to sink, and they are able to pass through sieve tube elements. But since they spread more slowly than solutes, for this and other reasons, movement by passive translocation is not supported.

## Detection and Diagnosis

Before molecular techniques were developed, the diagnosis of phytoplasma diseases was difficult because they could not be cultured. Thus, classical diagnostic techniques, such as observation of symptoms, were used. Ultrathin sections of the phloem tissue from suspected phytoplasma-infected plants would also be examined for their presence. Treating infected plants with antibiotics such as tetracycline to see if this cured the plant was another diagnostic technique employed.

Molecular diagnostic techniques for the detection of phytoplasma began to emerge in the 1980s and included ELISA-based methods. In the early 1990s, polymerase chain reaction-based methods were developed that were far more sensitive than those that used ELISA, and RFLP analysis allowed the accurate identification of different strains and species of phytoplasma.

More recently, techniques have been developed that allow for assessment of the level of infection. Both quantitative PCR and bioimaging have been shown to be effective methods of quantifying the titre of phytoplasmas within the plant.

## Control

Phytoplasmas are normally controlled by the breeding and planting of disease resistant varieties of crops (believed to the most economically viable option) and by the control of the insect vector.

Tissue culture can be used to produce clones of phytoplasma-infected plants that are healthy. The chances of gaining healthy plants in this manner can be enhanced by the use of cryotherapy, freezing the plant samples in liquid nitrogen, before using them for tissue culture.

Work has also been carried out investigating the effectiveness of plantibodies targeted against phytoplasmas.

Tetracyclines are bacteriostatic to phytoplasmas. However, without continuous use of the antibiotic, disease symptoms reappear. Thus, tetracycline is not a viable control agent in agriculture, but it is used to protect ornamental coconut trees.

## Genetics

The genomes of three phytoplasmas have been sequenced: aster yellows witches broom, onion yellows (*Ca*. Phytoplasma asteris) and *Ca*. Phytoplasma australiense Phytoplasmas have very small genomes, which also have extremely low levels of the nucleotides G and C, sometimes as little as 23%, which is thought to be the threshold for a viable genome. In fact Bermuda grass white leaf phytoplasma has a genome size of just 530 kb, one of the smallest known genomes of living organisms. Larger phytoplasma genomes are around 1350 kb. The small genome size associated with phytoplasmas is due to their being the product of reductive evolution from *Bacillus/Clostridium* ancestors. They have lost 75% or more of their original genes, so can no longer survive outside of insects or plant phloem. Some phytoplasmas contain extrachromo-somal DNA such as plasmids.

Despite their very small genomes, many predicted genes are present in multiple copies. Phyto-plasmas lack many genes for standard metabolic functions and have no functioning homologous recombination pathways, but do have a *sec* transport pathway. Many phytoplasmas contain two rRNA operons. Unlike the rest of the Mollicutes, the triplet code of UGA is used as a stop codon in phytoplasmas.

Phytoplasma genomes contain large numbers of transposon genes and insertion sequences. They also contain a unique family of repetitive extragenic palindromes called PhREPS whose role is un-known though it is theorised that the stem loop structures the PhREPS are capable of forming may play a role in transcription termination or genome stability.

## Taxonomy

Phytoplasmas belong to the monophyletic order Acholeplasmatales. In 1992, the Subcommittee on the Taxonomy of Mollicutes proposed the use of the name *Phytoplasma* in place of the use of the term MLO (mycoplasma-like organism) "for reference to the phytopathogenic mollicutes". In 2004, the genus name *Phytoplasma* was adopted and is currently at Candidatus status which is used for bacteria that cannot be cultured. Its taxonomy is complicated because it can not be cul-tured, thus methods normally used for classification of prokaryotes are not possible. Phytoplasma taxonomic groups are based on differences in the fragment sizes produced by the restriction digest of the 16S rRNA gene sequence (RFLP) or by comparison of DNA sequences from the 16s/23s spacer regions. There is some disagreement over how many taxonomic groups the phytoplasmas fall into, recent work involving computer simulated restriction digests of the 16Sr gene suggest there may be up to 28 groups whereas other papers argue for less groups, but more subgroups. Each group includes at least one *Ca*. Phytoplasma species, characterized by distinctive biological, phytopathological, and genetic properties.

# Magnaporthe Grisea

*Magnaporthe grisea*, also known as rice blast fungus, rice rotten neck, rice seedling blight, blast of rice, oval leaf spot of graminea, pitting disease, ryegrass blast, and Johnson spot, is a plant-patho-genic fungus that causes a serious disease affecting rice. It is now known that *M. grisea* consists of a cryptic species complex containing at least two biological species that have clear genetic differences and do not interbreed. Complex members isolated from *Digitaria* have been more narrowly defined as *M. grisea*. The remaining members of the complex isolated from rice and a variety of other hosts have been renamed *Magnaporthe oryzae*. Confusion on which of these two names to use for the rice blast pathogen remains, as both are now used by different authors.

Members of the *Magnaporthe grisea* complex can also infect other agriculturally important cereals including wheat, rye, barley, and pearl millet causing diseases called blast disease or blight disease. Rice blast causes economically significant crop losses annually. Each year it is estimated to destroy enough rice to feed more than 60 million people. The fungus is known to occur in 85 countries worldwide.

Hosts and symptoms

Lesions on rice leaves caused by infection with *M. grisea*

Rice blast lesions on plant nodes

*M. grisea* is an ascomycete fungus. It is an extremely effective plant pathogen as it can reproduce both sexually and asexually to produce specialized infectious structures known as appressoria that infect aerial tissues and hyphae that can infect root tissues.

Rice blast has been observed on rice strains M-201, M-202, M-204, M-205, M-103, M-104, S-102, L-204, Calmochi-101, with M-201 being the most vulnerable. Initial symptoms are white to gray-green lesions or spots with darker borders produced on all parts of the shoot, while older lesions are elliptical or spindle-shaped and whitish to gray with necrotic borders. Lesions may enlarge and coalesce to kill the entire leaf. Symptoms are observed on all above-ground parts of the plant. Lesions can be seen on the leaf collar, culm, culm nodes, and panicle neck node. Internodal infection of the culm occurs in a banded pattern. Nodal infection causes the culm to break at the infected node (rotten neck). It also affects reproduction by causing the host to produce fewer seeds. This is caused by the disease preventing maturation of the actual grain.

## Disease Cycle

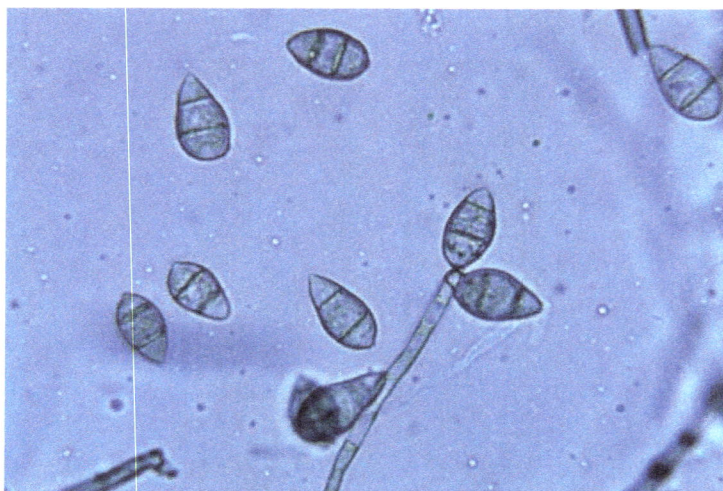

Spores of *M. grisea*

The pathogen infects as a spore that produces lesions or spots on parts of the rice plant such as the leaf, leaf collar, panicle, culm and culm nodes. Using a structure called an appressorium, the pathogen penetrates the plant. *M. grisea* then sporulates from the diseased rice tissue to be dispersed as conidiospores. After overwintering in sources such as rice straw and stubble, the cycle repeats.

A single cycle can be completed in about a week under favorable conditions where one lesion can generate up to thousands of spores in a single night. With the ability to continue to produce the spores for over 20 days, rice blast lesions can be devastating to susceptible rice crops.

## Environment

Rice blast is a significant problem in temperate regions and can be found in areas such as irrigated lowland and upland. Conditions conducive for rice blast include long periods of free moisture where leaf wetness is required for infection and high humidity is common. Sporulation increases with high relative humidity and at 77-82 degrees F, spore germination, lesion formation, and sporulation are at optimum levels.

In terms of control, excessive use of nitrogen fertilization as well as drought stress increase rice susceptibility to the pathogen as the plant is placed in a weakened state and its defenses are low. Extended drain periods also favor infection as they aerate the soil, converting ammonium to nitrate and thus causing stress to rice crops, as well.

## Management

The fungus has been able to establish resistance to both chemical treatments and genetic resistance in some types of rice developed by plant breeders. It is thought that the fungus can achieve this by genetic change through mutation. In order to most effectively control infection by *M. grisea*, an integrated management program should be implemented to avoid overuse of a single control method and fight against genetic resistance. For example, eliminating crop residue could reduce the occurrence of overwintering and discourage inoculation in subsequent seasons. Another strategy would be to plant resistant rice varieties that are not as susceptible to infection by *M. grisea*. Knowledge of the pathogenicity of *M. grisea* and its need for free moisture suggest other control strategies such as regulated irrigation and a combination of chemical treatments with different modes of action. Managing the amount of water supplied to the crops limits spore mobility thus dampening the opportunity for infection. Chemical controls such as Carpropamid have been shown to prevent penetration of the appressoria into rice epidermal cells, leaving the grain unaffected.

J. Sendra rice affected by Magnaporthe grisea.

## Importance

Rice blast is the most important disease concerning the rice crop in the world. Since rice is an important food source for much of the world, its effects have a broad range. It has been found in over 85 countries across the world and reached the United States in 1996. Every year the amount of crops lost to rice blast could feed 60 million people. Although there are some resistant strains of rice, the disease persists wherever rice is grown. The disease has never been eradicated from a region.

## Soybean Cyst Nematode

The soybean cyst nematode (SCN), *Heterodera glycines*, is a plant-parasitic nematode and a devastating pest of the soybean (*Glycine max*) worldwide. The nematode infects the roots of soybean, and the female nematode eventually becomes a cyst. Infection causes various symptoms that may include chlorosis of the leaves and stems, root necrosis, loss in seed yield and suppression of root and shoot growth. SCN has threatened the U.S. crop since the 1950s, reducing returns to soybean producers by $500 million each year and reducing yields by as much as 75 percent. It is also a significant problem in the soybean growing areas of South America and Asia.

The second-stage juvenile, or J2, nematode is the only life stage that can penetrate roots. (The first-stage juvenile occurs in the egg, and third- and fourth-stages occur in the roots). The J2 enters the root moving through the plant cells to the vascular tissue where it feeds. The J2 induces cell division in the root to form specialized feeding sites. As the nematode feeds, it swells. The female swells so much that her posterior end bursts out of the root and she becomes visible to the naked eye. In contrast, the adult male regains a wormlike shape, and he leaves the root in order to find and fertilize the large females. The female continues to feed as she lays 200 to 400 eggs in a yellow gelatinous matrix, forming an egg sac which remains inside her. She then dies and her cuticle hardens forming a cyst. The eggs may hatch when conditions in the soil are favorable, the larvae developing inside the cyst and the biological cycle repeating itself. There are usually three generations in the year. In the autumn or in unfavorable conditions, the cysts containing dormant larvae may remain intact in the soil for several years. Although soybean is the primary host of SCN, other legumes can also serve as hosts.

Segment of soybean root infected with soybean cyst nematode. Signs of infection are white to brown cysts filled with eggs that are attached to root surfaces.

## Pathology

The aboveground symptoms of SCN infection are not unique to SCN infection, and could be confused with nutrient deficiency, particularly iron deficiency, stress from drought, herbicide injury or another disease. The first signs of infection are groups of plants with yellowing leaves that have stunted growth. The pathogen may also be difficult to detect on the roots, since stunted roots are also a common symptom of stress or plant disease. Observation of adult females and cysts on the roots is the only accurate way to detect and diagnose SCN infection in the field.

## Distribution

The SCN is thought to be a native of Asia. It was first found in the United States in 1954 and spread with the expansion of soybean growing areas. SCN was also found in Colombia in the 1980s, and more recently in the major soybean producing areas in Argentina and Brazil. SCN has also been reported from Iran and Italy.

## Locations

- Africa: Egypt

- Asia: Iran (Golestan Province and Mazandaran Province), China (Hebei, Hubei, Heilongjiang, Henan, Jiangsu, Liaoning), Indonesia (Java), Korean peninsula, Japan, Taiwan (unconfirmed), Russia (Amur District in the Far East).

- North America: Canada (Ontario), USA (Alabama, Arkansas, Delaware, Florida, Georgia, Illinois, Indiana, Iowa, Kansas, Kentucky, Louisiana, Maryland, Minnesota, Michigan, Mississippi, Missouri, Nebraska, New Jersey, North Carolina, North Dakota, Ohio, Oklahoma, Pennsylvania, South Carolina, South Dakota, Tennessee, Texas, Virginia and Wisconsin).

- South America: Argentina (unconfirmed), Brazil (unconfirmed), Chile, Columbia, Ecuador.

## Control

Cultural practices, such as crop rotation and the use of resistant cultivars, are used to limit the damage due to SCN. Because SCN is an obligate parasite (requires a living host), a crop rotation involving non-host plants can decrease the population of SCN and has been shown to be an effective management tool. Plants that are already stressed are more susceptible to infection, so good cultural practices, like maintaining soil fertility, pH and moisture can reduce the severity of infection. Chemical control with nematicides is not normally used because the economic and environmental costs are prohibitive.

## References

- Dusenbery, David B. (2009). Living at Micro Scale, pp. 20–25. Harvard University Press, Cambridge, Mass. ISBN 978-0-674-03116-6.

- Lewin, Benjamin.; Krebs, Jocelyn E.; Kilpatrick, Stephen T.; Goldstein, Elliott S.; Lewin, Benjamin. Genes IX. (2011). Lewin's genes. Sudbury, Mass.: Jones and Bartlett. p. 23. ISBN 9780763766320.

- Brian W. J. Mahy, Marc H. V. Van Regenmortel (ed.). Desk Encyclopedia of Plant and Fungal Virology. Academic Press. pp. 71–81. ISBN 978-0123751485.

- Pommerville, Jeffrey C (2014). Fundamentals of Microbiology. Burlington, MA: Jones and Bartlett Learning. p. 482. ISBN 978-1-284-03968-9.

- "Disease found in Japanese Larch Trees in Ireland". Department of Agriculture, Food & the Marine. 17 August 2010. Retrieved 17 February 2014.

- The University of Waikato (March 25, 2014). "Bacterial DNA – the role of plasmids". Themes — Bacteria in biotech. Biotechnology Learning Hub. Retrieved 2014-09-03.

- Kurahasi, Yoshio (1997). "Biological Activity of Carpropamid (KTU 3616): A new fungicide for rice blast disease". Journal of Pesticide Science. Retrieved 2014-02-25.

# 3

# Various Disorders in Plants

This chapter provides the reader with a broad understanding on the various disorders found in plants. It particularly focuses on physiological plant disorder, which is dissimilar to plant diseases caused by pathogens. Altering environmental conditions can usually prevent physiological plant disorder. The major causes of physiological plant disorder discussed in this chapter are phytotoxicity, boron deficiency, calcium deficiency, magnesium deficiency and phosphorus deficiency.

## Physiological Plant Disorder

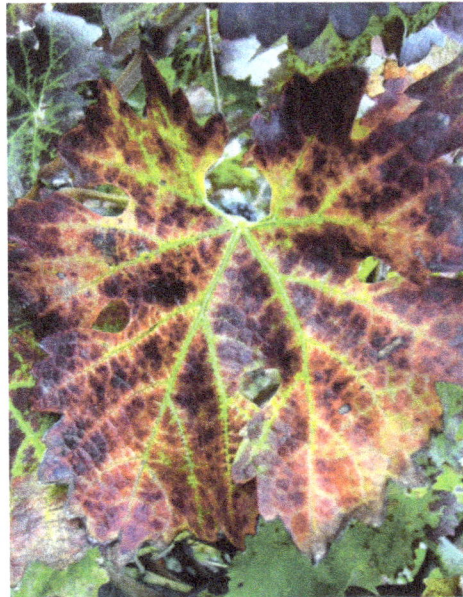

Deficit of micronutriens, vine.

Physiological plant disorders are caused by non-pathological conditions such as poor light, adverse weather, water-logging, phytotoxic compounds or a lack of nutrients, and affect the functioning of the plant system. Physiological disorders are distinguished from plant diseases caused by pathogens, such as a virus or fungus. While the symptoms of physiological disorders may appear disease-like, they can usually be prevented by altering environmental conditions. However, once a plant shows symptoms of a physiological disorder it is likely that that season's growth or yield will be reduced.

Diagnosis of the cause of a physiological disorder (or disease) can be difficult, but there are many web-based guides that may assist with this. Examples are: *Abiotic plant disorders: Symptoms, signs and solutions*; *Georgia Corn Diagnostic Guide*; *Diagnosing Plant Problems* (Kentucky); and *Diagnosing Plant Problems* (Virginia).

Sunburn at apple.

Some general tips to diagnosing plant disorders:

- Examine where symptoms first appear on a plant—on new leaves, old leaves or all over?

- Note the pattern of any discolouration or yellowing—is it all over, between the veins or around the edges? If only the veins are yellow deficiency is probably not involved.

- Note general patterns rather than looking at individual plants—are the symptoms distributed throughout a group of plants of the same type growing together. In the case of a deficiency all of the plants should be similarly effected, although distribution will depend on past treatments applied to the soil.

- Soil analysis, such as determining pH, can help to confirm the presence of physiological disorders.

- Consider recent conditions, such as heavy rains, dry spells, frosts, etc., may also help to determine the cause of plant disorders.

## Weather Damage

Frost and cold are major causes of crop damage to tender plants, although hardy plants can also suffer if new growth is exposed to a hard frost following a period of warm weather. Symptoms will often appear overnight, affecting many types of plants. Leaves and stems may turn black, and buds and flowers may be discoloured, and frosted blooms may not produce fruit. Many annual plants, or plants grown in frost free areas, can suffer from damage when the air temperature drops below 40 degrees Fahrenheit (4 degrees Celsius). Tropical plants may begin to experience cold damage when the temperature is 42 to 48 °F (5 to 9 °C), symptoms include wilting of the top of the stems and/or leaves, and blackening or softening of the plant tissue.

Frost or cold damage can be avoided by ensuring that tender plants are properly hardened before planting, and that they are not planted too early in the season, before the risk of frost has passed. Avoid planting susceptible plants in frost pockets, or where they will receive early morning sun. Protect young buds and bloom with horticultural fleece if frost is forecast. Cold, drying easterly winds can also severely inhibit spring growth even without an actual frost, thus adequate shelter or the use of windbreaks is important.

Drought can cause plants to suffer from water stress and wilt. Adequate irrigation is required during prolonged hot, dry periods. Rather than shallow daily watering, during a drought water should be directed towards the roots, ensuring that the soil is thoroughly soaked two or three times a week. Mulches also help preserve soil moisture and keep roots cool.

Heavy rains, particularly after prolonged dry periods, can also cause roots to split, onion saddleback (splitting at the base), tomatoes split and potatoes to become deformed or hollow. Using mulches or adding organic matter such as leaf mold, compost or well rotted manure to the soil will help to act as a 'buffer' between sudden changes in conditions. Water-logging can occur on poorly drained soils, particularly following heavy rains. Plants can become yellow and stunted, and will tend to be more prone to drought and diseases. Improving drainage will help to alleviate this problem.

Hail can cause damage to soft skinned fruits, and may also allow brown rot or other fungi to penetrate the plant. Brown spot markings or lines on one side of a mature apple are indicative of a spring hailstorm.

Plants affected by salt stress are unable to take water from soil, due to an osmotic imbalance between soil and plant.

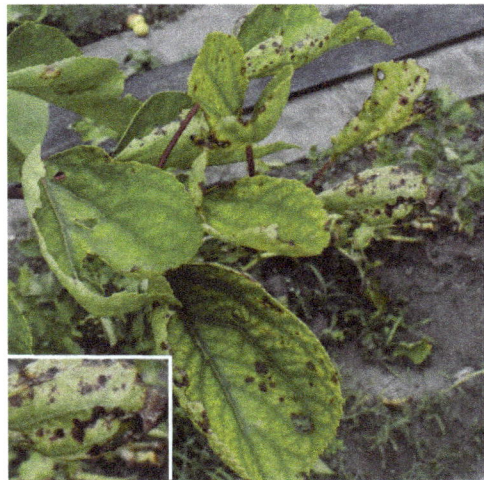

Iron deficiency.

## Nutrient Deficiencies

Poor growth and a variety of disorders such as leaf discolouration (chlorosis) can be caused by a shortage of one or more plant nutrients. Poor plant uptake of a nutrient from the soil (or other growing medium) may be due to an absolute shortage of that element in the growing medium, or because that element is present in a form that is not available to the plant. The latter can be caused by incorrect pH, shortage of water, poor root growth or an excess of another nutrient. Plant nutrient deficiencies can be avoided or corrected using a variety of approaches including the consultation of experts on-site, the use of soil and plant-tissue testing services, the application of prescription-blend fertilizers, the application of fresh or well-decomposed organic matter, and the use of biological systems such as cover crops, intercropping, improved fallows, ley cropping, permaculture, or crop rotation.

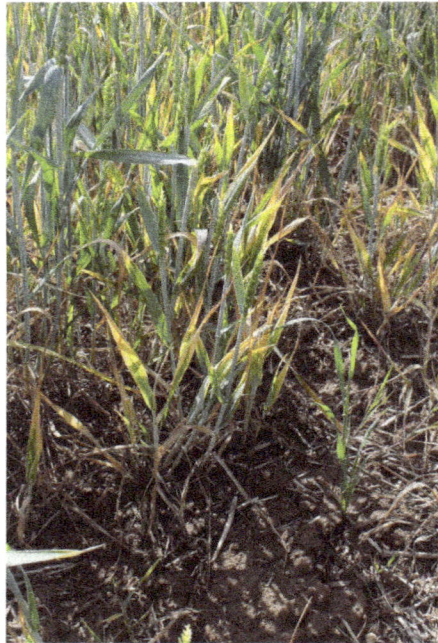

Drought.

Nutrient (or mineral) deficiencies include:

- Boron deficiency

- Calcium deficiency

- Iron deficiency

- Magnesium deficiency

- Manganese deficiency

- Nitrogen deficiency

- Phosphorus deficiency

- Potassium deficiency

- Zinc deficiency

- Shortage of trace elements such as molybdenum can also cause disorders such as whiptail in cauliflower.

## Cause of Physiological Plant Disorders

### Phytotoxicity

Phytotoxicity is a toxic effect by a compound on plant growth. Such damage may be caused by a wide variety of compounds, including trace metals, salinity, pesticides, phytotoxins or allelochemicals.

## Substances With Phytotoxic Potential

### Inorganic Compounds

High concentrations of mineral salts in solution within the growing medium can have phytotoxic effects. Sources of excessive mineral salts include infiltration of seawater and excessive application of fertilizers. For example urea is used in agriculture as a nitrogenous fertilizer, but if too much is applied, phytotoxic effects can result, either by urea toxicity or by the "ammonia produced through hydrolysis of urea by soil urease". Acid soils may contain high concentrations of aluminium (as $Al^{3+}$) and manganese (as $Mn^{2+}$) which can be phytotoxic.

### Herbicides

Herbicides are designed to kill plants, and are used to control unwanted plants such as agricultural weeds. However herbicides can also cause phytotoxic effects in plants that are not within the area over which the herbicide is applied, for example as a result of wind-blown spray drift or from the use of herbicide-contaminated material (such as straw or manure) being applied to the soil. The phytotoxic effects of herbicides are an important subject of study in the field of ecotoxicology.

### Boron Deficiency (Plant Disorder)

Boron deficiency is a common deficiency of the micronutrient boron in plants. It is the most widespread micronutrient deficiency around the world and causes large losses in crop production and crop quality. Boron deficiency affects vegetative and reproductive growth of plants, resulting in inhibition of cell expansion, death of meristem, and reduced fertility.

Plants contain boron both in a water-soluble and insoluble form. In intact plants, the amount of water-soluble boron fluctuates with the amount of boron supplied, while insoluble boron does not. The appearance of boron deficiency coincides with the decrease of water-insoluble boron. It appears that the insoluble boron is the functional form while the soluble boron represents the surplus.

Boron is essential for the growth of higher plants. The primary function of the element is to provide structural integrity to the cell wall in plants. Other functions likely include the maintenance of the plasma membrane and other metabolic pathways.

### Symptoms

Symptoms include dying growing tips and bushy stunted growth, extreme cases may prevent fruit set. Crop-specific symptoms include;

- *Apple*- interacting with calcium, may display as "water core", internal areas appearing frozen
- *Beetroot*- rough, cankered patches on roots, internal brown rot.
- *Cabbage*- distorted leaves, hollow areas in stems.
- *Cauliflower*- poor development of curds, and brown patches. Stems, leafstalks and midribs roughened.

- *Celery-* leaf stalks develop cracks on the upper surface, inner tissue is reddish brown.

- *Celeriac-* causes brown heart rot

- *Pears-* new shoots die back in spring, fruits develop hard brown flecks in the skin.

- *Strawberries-* Stunted growth, foliage small, yellow and puckered at tips. Fruits are small and pale.

- *Swede (rutabaga)* and *turnip-* brown or gray concentric rings develop inside the roots.

- *Arecaceae (Palm Tree)* - brown spots on fronds & lower productivity.

## Soil Conditions

Boron is present in the soil in many forms, the most common being Boric Acid ($H_3BO_3$). An adequate amount of boron in the soil is 12 mg/kg. If the boron content of the soil drops below 0.14 mg/kg then boron deficiency is likely to be observed. Boron deficiency is also observed in basic soils with a high pH because in basic conditions boric acid exists in an undissociated form which the plant is unable to absorb. Soils with low organic matter content (<1.5%) are also susceptible to boron deficiency. Highly leached sandy soils are also characteristic of boron deficiency because the boron will not be retained in the soil. Boron toxicity is also possible if the boron content of the soil is high enough that the plant cannot cope with the excess boron. The levels at which boron is toxic to plants varies with different species of plants.

## Boron Requirements

Boron is an essential micronutrient which means it is essential for plant growth and development, but is required in very small quantities. Although Boron requirements vary among crops, the optimum boron content of the leaves for most crops is 20-100 ppm. Excess boron can result in boron toxicity and the toxicity level varies between plants.

## Treatment

Boric acid (16.5%boron), borax (11.3% boron) or SoluBor (20.5% boron) can be applied to soils to correct boron deficiency. Typical applications of actual boron are about 1.1 kg/hectare or 1.0 lb/acre but optimum levels of boron vary with plant type. Borax, Boric Acid or Solubor can be dissolved in water and sprayed or applied to soil as a dust. Excess boron is toxic to plants so care must be taken to ensure correct application rate and even coverage. Leaves of many plants are damaged by boron; therefore, when in doubt, only apply to soil. Application of boron may not correct boron deficiency in alkaline soils because even with the addition of boron, it may remain unavailable for plant absorption. Continued application of boron may be necessary in soils that are susceptible to leaching such as sandy soils. Flushing soils containing toxic levels of boron with water can remove the boron through leaching.

## Functions

Once boron has been absorbed by the plant and incorporated into the various structures that require boron, it is unable to disassemble these structures and re-transport boron through the plant resulting in boron being a non-mobile nutrient. Due to translocation difficulties the youngest leaves often show deficiency symptoms first.

## Cell Wall

Boron is part of the dRG-II-B complex which is involved in the cross linking for pectin located in the primary cell wall and the middle lamella of plant cells. This cross linking is thought to stabilize the matrix of plant cell walls.

## Carbohydrate Metabolism

## Protein Synthesis

## Germination and Pollination

The B requirement is much higher for reproductive growth than for vegetative growth in most plant species. Boron increases flower production and retention, pollen tube elongation and germination, and seed and fruit development.

A deficiency of B can cause incomplete pollination of corn or prevent maximum pod-set in soybeans.

## Sugar Translocation

Photosynthesis transforms sunlight energy into plant energy compounds such as sugars. For this process to continue in plants, the sugars must be moved away from the site of their development, and stored or used to make other compounds.

Boron increases the rate of transport of sugars (which are produced by photosynthesis in mature plant leaves) to actively growing regions and also in developing fruits. Boron is essential for providing sugars which are needed for root growth in all plants and also for normal development of root nodules in legumes such as alfalfa, soybeans and peanuts.

## Calcium Deficiency (Plant Disorder)

Calcium deficiency (blossom end rot) on a tomato

Calcium (Ca) deficiency is a plant disorder that can be caused by insufficient calcium in the growing medium, but is more frequently a product of low transpiration of the whole plant or more commonly the affected tissue. Plants are susceptible to such localized calcium deficiencies in low or nontranspiring tissues because calcium is not transported in the phloem. This may be due to water shortages, which slow the transportation of calcium to the plant, poor uptake of calcium through the stem, or too much nitrogen in the soil.

## Causes

Acidic, sandy, or coarse soils often contain less calcium. Uneven soil moisture and overuse of fertilizers can also cause calcium deficiency. At times, even with sufficient calcium in the soil, it can be in an insoluble form and is then unusable by the plant or it could be attributed to a "transport protein". Soils containing high phosphorus are particularly susceptible to creating insoluble forms of calcium.

## Symptoms

Blossom end rot on a grape tomato

Calcium deficiency symptoms appear initially as localized tissue necrosis leading to stunted plant growth, necrotic leaf margins on young leaves or curling of the leaves, and eventual death of terminal buds and root tips. Generally, the new growth and rapidly growing tissues of the plant are affected first. The mature leaves are rarely if ever affected because calcium accumulates to high concentrations in older leaves.

Crop-specific symptoms include:

Apple

     'Bitter pit' – fruit skins develop pits, brown spots appear on skin and/or in flesh and taste of those areas is bitter. This usually occurs when fruit is in storage, and Bramley apples are particularly susceptible. Related to boron deficiency, "water cored" apples seldom display bitter pit effects.

Cabbage and Brussels sprouts

Internal browning and "tip burn"

Carrot

'Cavity spot' – oval spots develop into craters which may be invaded by other disease-causing organisms.

Celery

Stunted growth, central leaves stunted.

Tomatoes and peppers

'Blossom end rot' – Symptoms start as sunken, dry decaying areas at the blossom end of the fruit, furthest away from the stem, not all fruit on a truss is necessarily affected. Sometimes rapid growth from high-nitrogen fertilizers may exacerbate blossom end rot.

## Treatment

Dissection of grape tomato with blossom end rot

Calcium deficiency can sometimes be rectified by adding agricultural lime to acid soils, aiming at a pH of 6.5, unless the subject plants specifically prefer acidic soil. Organic matter should be added to the soil to improve its moisture-retaining capacity. However, because of the nature of the disorder (i.e. poor transport of calcium to low transpiring tissues), the problem cannot generally be cured by the addition of calcium to the roots. In some species, the problem can be reduced by prophylactic spraying with calcium chloride of tissues at risk.

Plant damage is difficult to reverse, so corrective action should be taken immediately, supplemental applications of calcium nitrate at 200 ppm nitrogen, for example. Soil pH should be tested, and corrected if needed, because calcium deficiency is often associated with low pH.

## Magnesium Deficiency (Plants)

Magnesium (Mg) deficiency is a detrimental plant disorder that occurs most often in strongly acidic, light, sandy soils, where magnesium can be easily leached away. Magnesium is an essential micro nutrient found from 0.2-0.4% dry matter and is necessary for normal plant growth.

A plant with Magnesium deficiency

Excess potassium, generally due to fertilizers, further aggravates the stress from the magnesium deficiency, as does aluminium toxicity.

Magnesium has an important role in photosynthesis because it forms the central atom of chlorophyll. Therefore, without sufficient amounts of magnesium, plants begin to degrade the chlorophyll in the old leaves. This causes the main symptom of magnesium deficiency, chlorosis, or yellowing between leaf veins, which stay green, giving the leaves a marbled appearance. Due to magnesium's mobile nature, the plant will first break down chlorophyll in older leaves and transport the Mg to younger leaves which have greater photosynthetic needs. Therefore, the first sign of magnesium deficiency is the chlorosis of old leaves which progresses to the young leaves as the deficiency continues. Magnesium also is a necessary activator for many critical enzymes, including ribulosbiphosphate carboxylase (RuBisCO) and phosphoenolpyruvate carboxylase (PEPC), both essential enzymes in carbon fixation. Thus low amounts of Mg lead to a decrease in photosynthetic and enzymatic activity within the plants. Magnesium is also crucial in stabilizing ribosome structures, hence, a lack of magnesium causes depolymerization of ribosomes leading to pre-mature aging of the plant. After prolonged magnesium deficiency, necrosis and dropping of older leaves occurs. Plants deficient in magnesium also produce smaller, woodier fruits.

Magnesium deficiency may be confused with zinc or chlorine deficiencies, viruses, or natural aging since all have similar symptoms. Adding Epsom salts (as a solution of 25 grams per liter or 4 oz per gal) or crushed dolomitic limestone to the soil can rectify magnesium deficiencies. For a more organic solution, applying home-made compost mulch can prevent leaching during excessive rainfall and provide plants with sufficient amounts of nutrients, including magnesium.

## Nitrogen Deficiency

A young cabbage plant exhibiting nitrogen deficiency.

All plants require sufficient supplies of macronutrients for healthy growth, and nitrogen (N) is a nutrient that is commonly in limited supply. Nitrogen deficiency in plants can occur when organic matter with high carbon content, such as sawdust, is added to soil. Soil organisms use any nitrogen to break down carbon sources, making N unavailable to plants. This is known as "robbing" the soil of nitrogen. All vegetables apart from nitrogen fixing legumes are prone to this disorder.

Nitrogen deficiency can be prevented in the short term by using grass mowings as a mulch, or foliar feeding with manure, and in the longer term by building up levels of organic matter in the soil. Sowing green manure crops such as grazing rye to cover soil over the winter will help to prevent nitrogen leaching, while leguminous green manures such as winter tares will fix additional nitrogen from the atmosphere.

## Symptoms

Some symptoms of nitrogen deficiency (in absence or low supply) are given below :

1. The chlorophyll content of the plant leaves is reduced which results in pale yellow colour. Older leaves turn completely yellow.

2. Flowering, fruitings, protein and starch contents are reduced. Reduction in protein results in stunted growth and dormant lateral buds.

## Disease

Plants look thin, pale and the condition is called *general starvation*.

## Detection

The visual symptoms of nitrogen deficiency mean that it can be relatively easy to detect in some plant species. Symptoms include poor plant growth, and leaves that are pale green or yellow because they are unable to make sufficient chlorophyll. Leaves in this state are said to be chlorotic. Lower leaves (older leaves) show symptoms first, since the plant will move nitrogen from older tissues to more important younger ones. Nevertheless, plants are reported to show nitrogen deficiency symptoms at different parts. For example, Nitrogen deficiency of tea is identified by retarded shoot growth and yellowing of younger leaves.

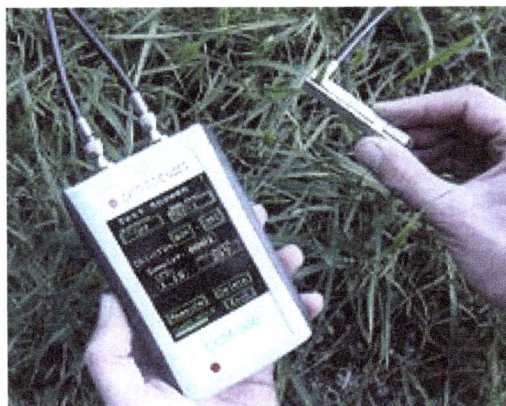

By measuring chlorophyll content Nitrogen deficiency can be detected.

However, these physical symptoms can also be caused by numerous other stresses, such as deficiencies in other nutrients, toxicity, herbicide injury, disease, insect damage or environmental conditions. Therefore, nitrogen deficiency is most reliably detected by conducting quantitative tests in addition to assessing the plants visual symptoms. These tests include soil tests and plant tissue test.

Plant tissue tests destructively sample the plant of interest. However, nitrogen deficiency can also be detected non-destructively by measuring chlorophyll content.

Chlorophyll content tests work because leaf nitrogen content and chlorophyll concentration are closely linked, which would be expected since the majority of leaf nitrogen is contained in chlorophyll molecules. Chlorophyll content can be detected with a Chlorophyll content meter; a portable instrument that measures the greenness of leaves to estimate their relative chlorophyll concentration.

Chlorophyll content can also be assessed with a chlorophyll fluorometer, which measures a chlorophyll fluorescence ratio to identify phenolic compounds that are produced in higher quantities when nitrogen is limited. These instruments can therefore be used to non-destructively test for nitrogen deficiency.

## Corrective Measures

Fertilisers like ammonium phosphate, calcium ammonium nitrate, urea can be supplied. Foliar spray of urea can be a quick method.

## Phosphorus Deficiency

Phosphorus deficiency is a plant disorder associated with insufficient supply of phosphorus. Phosphorus refers here to salts of phosphates ($PO_4^{3-}$), monohydrogen phosphate ($HPO_4^{2-}$), and dihydrogen phosphate ($H_2PO_4^-$). These anions readily interconvert, and the predominant species is determined by the pH of the solution or soil. Phosphates are required for the biosynthesis of genetic material as well as ATP, essential for life. Phosphorus deficiency can be controlled by applying sources of phosphorus such as bone meal, rock phosphate, manure, and phosphate-fertilizers.

Phosphorus deficiency on corn

## Symptoms (Biological Implications)

In plants, Phosphorus (P) is considered second to nitrogen as the most essential nutrient to ensure health and function. Phosphorus is used by plants in numerous processes such as photophosphorylation, genetic transfer, the transportation of nutrients, and phospholipid cell membranes. Within a plant cell these functions are imperative for function, in photophosphoroylation for example the creation of stored energy in plants is a result of a chemical reaction including phosphorus. Phosphorus is a key molecular component of genetic reproduction. When phosphorus is present in inadequate levels, genetic processes such as cell division and plant growth are impaired. Hence, phosphorus deficient plants may mature at a slower rate than plants with adequate amounts of phosphorus. The stunted growth induced by phosphorus deficiency has been correlated with smaller leaf sizes and a lessened number of leaves. Phosphorus deficiency may also create an imbalance in the storage of carbohydrates. Photosynthesis, the main function of plant cells that produces energy from sunlight and water, usually remains at a normal rate under a phosphorus-deficient state. However phosphorus usage in functions within the cell usually slow. This imbalance of rates in phosphorus deficient plants leads to the buildup of excess carbohydrate within the plant. This carbohydrate buildup often can be observed by the darkening of leaves. In some plants the leaf pigment change as a result of this process can turn leaves a dark purplish color.

## Detection

Detecting phosphorus deficiency can take multiple forms. A preliminary detection method is a visual inspection of plants. Darker green leaves and purplish or red pigment can indicate a deficiency in phosphorus. This method however can be an unclear diagnosis because other plant environment factors can result in similar discoloration symptoms. In commercial or well monitored settings for plants, phosphorus deficiency is diagnosed by scientific testing. Additionally, discoloration in plant leaves only occurs under fairly severe phosphorus deficiency so it is beneficial to planters and farmers to scientifically check phosphorus levels before discoloration occurs. The most prominent method of checking phosphorus levels is by soil testing. The major soil testing methods are Bray 1-P, Mehlich 3, and Olsen methods. Each of these methods are viable but each method has tendencies to be more accurate in known geographical areas. These tests use chemical solutions to extract phosphorus from the soil. The extract must then be analyzed to determine the concentration of the phosphorus. Colorimetry is used to determine this concentration. With the addition of the phosphorus extract into a colorimeter, there is visual color change of the solution and the degree to this color change is an indicator of phosphorus concentration. To apply this testing method on phosphorus deficiency, the measured phosphorus concentration must be compared to known values. Most plants have established and thoroughly tested optimal soil conditions. If the concentration of phosphorus measured from the colorimeter test is significantly lower than the plant's optimal soil levels, then it is likely the plant is phosphorus deficient. The soil testing with colorimetric analysis, while widely used, can be subject to diagnostic problems as a result of interference from other present compounds and elements. Additional phosphorus detection methods such as spectral radiance and inductively coupled plasma spectrometry (ICP) are also implemented with the goal of improving reading accuracy. According to the World Congress of Soil Scientists, the advantages of these light-based measurement methods are their quickness of evaluation, simultaneous measurements of plant nutrients, and their non-destructive testing nature. Although these methods have experimental based evidence, unanimous approval of the methods has not yet been achieved.

## Treatment

Correction and prevention of phosphorus deficiency typically involves increasing the levels of available phosphorus into the soil. Planters introduce more phosphorus into the soil with bone meal, rock phosphate, manure, and phosphate-fertilizers. The introduction of these compounds into the soil however does not ensure the alleviation of phosphorus deficiency. There must be phosphorus in the soil, but the phosphorus must also be absorbed by the plant. The uptake of phosphorus is limited by the chemical form in which the phosphorus is available in the soil. A large percentage of phosphorus in soil is present in chemical compounds that plants are incapable of absorbing. Phosphorus must be present in soil in specific chemical arrangements to be usable as plant nutrients. Facilitation of usable phosphorus in soil can be optimized by maintaining soil within a specified pH range. Soil acidity, measured on the pH scale, partially dictates what chemical arrangements that phosphorus forms. Between pH 6 and 7, phosphorus makes the fewest number of bonds which render the nutrient unusable to plants. At this range of acidity the likeliness of phosphorus uptake is increased and the likeliness of phosphorus deficiency is decreased. Another component in the prevention and treatment of phosphorus is the plant's disposition to absorb nutrients. Plant species and different plants within in the same species react differently to low levels of phosphorus in soil. Greater expansion of root systems generally correlate to greater nutrient uptake. Plants within a species that have larger roots are genetically advantaged and less prone to phosphorus deficiency. These plants can be cultivated and bred as a long term phosphorus deficiency prevention method. In conjunction to root size, other genetic root adaptations to low phosphorus conditions such as mycorrhizal symbioses have been found to increase nutrient intake. These biological adaptations to roots work to maintain the levels of vital nutrients. In larger commercial agriculture settings, variation of plants to adopt these desirable phosphorus intake adaptations may be a long-term phosphorus deficiency correction method.

## Zinc Deficiency (Plant Disorder)

Maize plants with severe zinc deficiency in the foreground, with healthier plants (planted at the same time) in the background.

Zinc deficiency occurs when plant growth is limited because the plant cannot take up sufficient quantities of this essential micronutrient from its growing medium. It is one of the most widespread micronutrient deficiencies in crops and pastures worldwide and causes large losses in crop production and crop quality.

Almost half of the world's cereal crops are grown on zinc-deficient soils; as a result, zinc deficiency in humans is a widespread problem.

## Symptoms

Visible deficiency symptoms include:

- Chlorosis - yellowing of leaves; often interveinal; in some species, young leaves are the most affected, but in others both old and new leaves are chlorotic;

- Necrotic spots - death of leaf tissue on areas of chlorosis;

- Bronzing of leaves - chlorotic areas may turn bronze coloured;

- Rosetting of leaves - zinc-deficient dicotyledons often have shortened internodes, so leaves are clustered on the stem;

- Stunting of plants - small plants may occur as a result of reduced growth or because of reduced internode elongation;

- Dwarf leaves ('little leaf') - small leaves that often show chlorosis, necrotic spots or bronzing;

- Malformed leaves - leaves are often narrower or have wavy margins.

Macadamia shoots showing zinc deficiency symptoms. The youngest leaves show chlorosis (yellowing), dwarfing and malformation.

## Soil Conditions

Zinc deficiency is common in many different types of soil; some soils (sandy soils, histosols and soils developed from highly weathered parent material) have low total zinc concentrations, and others have low plant-available zinc due to strong zinc sorption (calcareous soils, highly weathered soils, vertisols, hydromorphic soils, saline soils). Soils low in organic matter (such as where topsoils have been removed), and compacted soils that restrict root proliferation also have a high risk of zinc deficiency. Application of phosphorus fertilizers has frequently been associated with zinc deficiency; this may be due to enhanced sorption by clay minerals (especially iron oxides), suppression of vesicular arbuscular mycorrhizae and/or immobilization of zinc in plant tissues. Liming of soils also frequently induces zinc deficiency by increasing zinc sorption.

## Zinc Requirements

Zinc is an essential micronutrient which means it is essential for plant growth and development, but is required in very small quantities. Although zinc requirements vary among crops, zinc leaf concentrations (on a dry matter basis) in the range 20 to 100 mg/kg are adequate for most crops.

## Treatment

Zinc sulphate or zinc oxide can be applied to soils to correct zinc deficiency. Recommended applications of actual zinc range from 5 to 100 kg/hectare but optimum levels of zinc vary with plant type and the severity of the deficiency. Application of zinc may not correct zinc deficiency in alkaline soils because even with the addition of zinc, it may remain unavailable for plant absorption.

Foliar applications of zinc as zinc sulphate or as zinc chelate (or other organic complexes) are also widely used, especially with fruit trees and grape vines. Zinc can also be supplied as a seed treatment, or by root-dipping of transplant seedlings.

## Functions

Zinc occurs in plants as a free ion, as a complex with a variety of low molecular weight compounds, or as a component of proteins and other macromolecules. In many enzymes, zinc acts as a functional, structural, or regulatory cofactor; a large number of zinc-deficiency disorders are associated with the disruption of normal enzyme activity (including that of key photosynthetic enzymes). Zinc deficiency also increases membrane leakiness as zinc-containing enzymes are involved in the detoxification of membrane-damaging oxygen radicals. Zinc may be involved in the control of gene expression; it appears important in stabilizing RNA and DNA structure, in maintaining the activity of DNA-synthesizing enzymes and in controlling the activity of RNA-degrading enzymes.

## References

- Mengel, K.; Kirkby, E.A. (2001). Principles of plant nutrition (5th ed.). Dordrecht: Kluwer Academic Publishers. ISBN 079237150X.

- Stefan Buczacki, Keith Harris (1998). "Disorders". Pests, Diseases & Disorders Of Garden Plants. Collins. p. 609. ISBN 0 00 220063 5.

- Norman P.A. Huner; William Hopkins. "3 & 4". Introduction to Plant Physiology 4th Edition. John Wiley & Sons, Inc. ISBN 978-0-470-24766-2.

- Merhaut, D.J. (2006). "Magnesium". In Barker A.V.; Pilbeam D.J. Handbook of plant nutrition. Boca Raton: CRC Press. p. 154. ISBN 9780824759049.

- Pandey, S N; Sinha, B K. "Mineral Nutrition". Plant Physiology (fourth ed.). 576Masjid Road, Jangpura, New Delhi-110014: VIKAS PUBLISHING HOUSE Pvt. Ltd. pp. 125–126. ISBN 8125918795.

- Alloway, B.J. (2008). Zinc in soils and crop nutrition (PDF). Brussels: International Zinc Association and International Fertilizer Industry Association. ISBN 9789081333108. Retrieved 14 April 2015.

- Weir, R.G,; Cresswell, G.C.; Loebel, M.R. (1995). Plant nutrient disorders 2: Tropical fruit and nut crops. Melbourne: Inkata Press. ISBN 0909605904.

- Lee, R.D. (2012). "Georgia Corn Diagnostic Guide". extension.uga.edu. University of Georgia Cooperative Extension. Retrieved 10 April 2015.

- Niemiera, A.X. (2009). "Diagnosing Plant Problems". pubs.ext.vt.edu. Virginia Cooperative Extension. Retrieved 10 April 2015.

- Camacho-Cristóbal, Juan J.; Jesús Rexach; Agustín González-Fontes. "Boron in plants: deficiency and toxicity" (PDF). Journal of Integrative Plant Science. Retrieved 2012-11-21.

# Plant Diseases: An Overview

This chapter will provide an integrated understanding of plant diseases. It offers an insightful focus, keeping in mind the complex subject matter. Organisms of various kinds cause these disorders. Phyllody, citrus canker, powdery scab, wilt disease, top dying disease and barley yellow dwarf are broadly explained in this chapter.

## Phyllody

Phyllody is the abnormal development of floral parts into leafy structures. It is generally caused by phytoplasma or virus infections, though it may also be because of environmental factors that result in an imbalance in plant hormones. Phyllody causes the affected plant to become partially or entirely sterile, as it is unable to normally produce flowers.

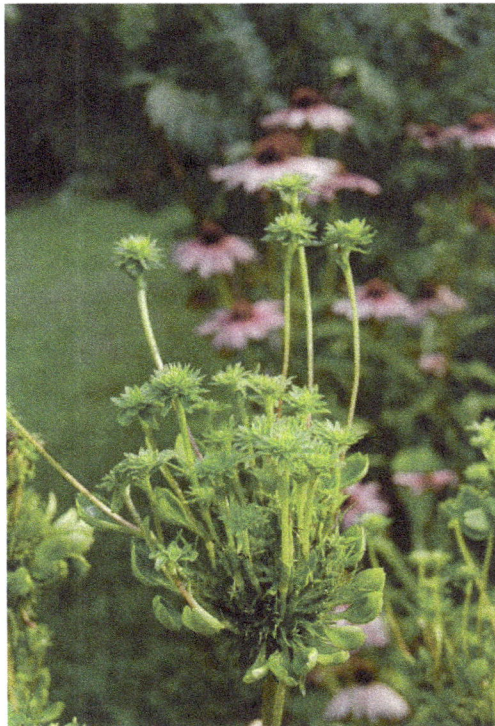

Phyllody on a purple coneflower (*Echinacea purpurea*)

The condition is also known as phyllomorphy or frondescence; though the latter may sometimes refer more generically to foliage, leafiness, or the process of leaf growth. Phyllody is usually differentiated from floral virescence, wherein the flowers merely turn green in color, but otherwise retain their normal structure. However, floral virescence and phyllody (along with witch's broom and other growth abnormalities), commonly occur together as symptoms of the same diseases. The term chloranthy is also often used for phyllody (particularly flowers exhibiting complete phyllody, such that it resembles leaf buds more than flowers), though in some cases it may refer to floral virescence.

## History

In the late 18th century, the German poet and philosopher Johann Wolfgang von Goethe noted strange-looking rose flowers where the flower organs were replaced by leafy or stem-like structures. This led him to hypothesize that plant organs arising from the stem are simply modifications of the same basic leaf organ. During growth, these organs naturally differentiate into specialized or generalized structures like petals or leaves. However, if certain factors interfere during the early growth stages, these organs can develop into something other than the original "plan of construction". He called this abnormal growth "metamorphosis" and it is the main topic of his essay *Versuch die Metamorphose der Pflanzen zu erklären* (1790), better known in English as the *Metamorphosis of Plants*. Goethe's hypothesis was poorly received by other scientists during his time, but it is now known to be essentially correct. The concepts he discusses while describing metamorphosis is now known as homology, the basis of the modern science of comparative anatomy and a discovery that is usually credited to the English biologist Sir Richard Owen.

In 1832, the German-American botanist George Engelmann described the same condition in his work *De Antholysi Prodromus*. He gave it the name "frondescence". Nineteen years later, the Belgian botanist Charles Jacques Édouard Morren also investigated the phenomenon in his book *Lobelia* (1851). Morren called the condition "phyllomorphy", and unlike Engelmann, Morren explicitly distinguished phyllomorphy (wherein the floral parts are replaced by leaf-like structures) from virescence (wherein the affected parts, not necessarily floral, turn green but retain the original form or structure).

The term "phyllody" was coined by the English botanist Maxwell T. Masters in his book on plant abnormalities, *Vegetable Teratology* (1869). The term is derived from Scientific Latin *phyllodium*, which is itself derived from Ancient Greek (*phullodes*, 'leaf-like'). Like Morren, Masters also distinguished phyllody from virescence. He acknowledged "frondescence" and "phyllomorphy" as synonyms of phyllody.

## Description

Phyllody is characterized by the partial or complete replacement of floral organs with true leaves. Phyllody can affect bracts, the calyx (sepals), corolla (petals), the gynoecium (carpels/pistils), and the androecium (stamens). Phyllody may be partial, affecting only some sets of floral organs or even only half of a set of floral organs (e.g. only three petals out of six in a single flower); or it can be complete, with all the floral organs replaced by leaves.

Illustration from *Vegetable Teratology* (1869), showing a *Petunia* flower with stamens partially replaced by "stalked" leaves. The stalks are actually the retained filaments of the stamens, while the anthers have been replaced by small leaves.

Longitudinal section of a rose flower exhibiting phyllody. Despite the apparent hips, the reproductive organs are completely absent and have been replaced by leaves.

Longitudinal section of a normal developing rose hip

Phyllody of the bracts is common among plants which bear catkin (amentaceous) inflorescences. They are very common among members of the genus *Plantago*, for example, as well as the common hop (*Humulus lupulus*). Involucral bracts of the flowers of members of the family Asteraceae like dahlias and dandelions, may also be affected.

Sepals that exhibit phyllody are usually hard to detect due to fact that most sepals already resemble leaves. Close examination, however, can reveal differences in venation in normal sepals and sepals that exhibit phyllody. The full development of perfect leaves from sepals is more common among flowers that have united sepals (monosepalous) than in flowers with separated sepals (polysepalous).

Phyllody of the petals can be expressed more mildly as a simple change in shape and color (in which case, it's more accurately virescence), or it can be expressed as fully formed leaves. It is more common among flowers which exhibit corollas of distinct petals (polypetalous) than in flowers in which the petals are fused into a single tube or bowl-like structure (monopetalous).

Phyllody of the stamens is rare. In fact, the stamens are the least likely of the floral organs to be affected by phyllody. This is thought to be because the stamens are the most highly differentiated organs in flowers.

In contrast, phyllody of the carpels is much more common than the corresponding changes in stamens. Usually, phyllody affects the proximal parts of the carpel (the ovary) more than the distal parts (the style and stigma). The ovule itself may be exposed on the edges or on the inner surface of the carpel if the ovary becomes leaf-like. If the ovule is affected by phyllody, it develops separately from the rest of the carpel. The best known example of phyllody of the carpels is found in the Japanese cherry (*Prunus serrulata*), in which one or both of the carpels can become leaf-like (although the distal half of the style and the stigma are usually unaffected). Incidentally, some Japanese cherry cultivars also exhibit "doubling" of the petals due to petalody, where a second corolla develops instead of stamens.

## Causes

## Biotic

In many cultivated plants, phyllody is caused by infections of plant pathogens and/or infestations of ectoparasites. Aside from exhibiting phyllody, they may also exhibit other symptoms like virescence, witch's brooms, chlorosis, and stunted growth. Examples of these biotic factors include:

- Phytoplasmas - specialized prokaryotic microorganisms that cause more than 200 distinct plant diseases. They resemble other bacteria but lack cell walls and are filamentous or pleomorphic in form. They are obligate parasites of plant phloem tissue and are spread by insect vectors. They are the most common cause of phyllody. Evidence suggests that phytoplasmas downregulate a gene involved in petal formation, instead causing leaves or leaflike structures to form. Examples of commercially important phytoplasma diseases are aster yellows, apple proliferation, clover phyllody, and *Sesamum* phyllody.

- Viruses, like the rose rosette disease (RRD)

- Fungi, like the smut fungus *Sphacelotheca reiliana* of corn and the rust fungus *Atelocauda koae* which infects *Acacia koa*

- Water molds, like *Sclerophthora macrospora* which infects more than 140 kinds of cereals including rice, corn, and wheat. The disease is more commonly known as "crazy top" because its most striking symptom is phyllody of the ears and tassels.

- Insect damage

In addition to causing phyllody itself, insects and other ectoparasites also serve as disease vectors that can spread phyllody to other nearby plants. The most common of these insect vectors are leafhoppers, an example of which is *Hishimonus phycitis*, which transmits the phytoplasma-caused little leaf phyllody in eggplants. The broken-backed bug (*Taylorilygus apicalis*) is another insect vector of a phytoplasma-caused phyllody in species of *Parthenium*. Other ectoparasite vectors include eriophyid mites, like the rose leaf curl mite (*Phyllocoptes fructiplilus*) which is known to be the primary vector of the rose rosette disease; and the chrysanthemum rust mite (*Paraphytoptus chrysanthemi*) which transmits phytoplasma-caused phyllody in species of chrysanthemums.

## Abiotic

Environmental abiotic factors like hot weather or water stress that result in an imbalance in plant hormones during flowering can cause phyllody. These can usually be differentiated from phyllody caused by biotic factors by the simultaneous presence of healthy and abnormal flowers. When conditions normalize, the plants resume normal flowering. The susceptibility of plants to environmentally caused phyllody can be genetic.

Phyllody in the green rose (*Rosa chinensis* var. *viridiflora*)

## Artificial

Phyllody can be artificially induced by applying cytokinins (CK), plant hormones responsible for cell division, as well as apical dominance and axillary bud growth. Conversely, it can be subsequently suppressed with the application of gibberellins (GA), plant hormones responsible for stem elongation, flowering, and sex expression.

## Related Floral Teratology

Other related floral development abnormalities are:

- Petalody - The transformation of floral organs (usually the stamens) into petals.
- Pistillody - The transformation of floral organs into pistils.
- Sepalody - The transformation of floral organs into sepals or sepal-like bodies.
- Staminody - The transformation of floral organs into stamens.

## Phyllody in Plant Breeding

In some cases, the occurrence of phyllody has been utilized in plant breeding. One of the most well known examples is the green rose (*Rosa chinensis* var. *viridiflora*), an ancient Chinese rose cultivar which exhibits green leafy bracts in tight flower-like clusters. In green rose, artificial selection has enabled phyllody to be expressed as a stable mutation.

# Citrus Canker

Citrus canker is a disease affecting *Citrus* species caused by the bacterium *Xanthomonas axonopodis*. Infection causes lesions on the leaves, stems, and fruit of citrus trees, including lime, oranges, and grapefruit. While not harmful to humans, canker significantly affects the vitality of citrus trees, causing leaves and fruit to drop prematurely; a fruit infected with canker is safe to eat, but too unsightly to be sold.

The disease, which is believed to have originated in Southeast Asia, is extremely persistent when it becomes established in an area. Citrus groves have been destroyed in attempts to eradicate the disease. Brazil and the United States are currently suffering from canker outbreaks.

## Biology

*Xanthomonas axonopodis* is a rod-shaped Gram-negative bacterium with polar flagella. The bacterium has a genome length around 5 megabase pairs. A number of types of citrus canker diseases are caused by different pathovars and variants of the bacterium:

- The Asiatic type of canker (canker A), *X. axonopodis* pv. *citri*, caused by a group of strains originally found in Asia, is the most widespread and severe form of the disease.

- Cancrosis B, caused by a group of *X. axonopodis* pv. *aurantifolii* strains originally found in South America is a disease of lemons, key lime, bitter orange, and pomelo.

- Cancrosis C, also caused by strains within *X. axonopodis* pv. *aurantifolii*, only infects key lime and bitter orange.

- A* strains, discovered in Oman, Saudi Arabia, Iran, and India, only infect key lime.

## Pathology

Plants infected with citrus canker have characteristic lesions on leaves, stems, and fruit with raised, brown, water-soaked margins, usually with a yellow halo or ring effect around the lesion. Older lesions have a corky appearance, still in many cases retaining the halo effect. The bacterium propagates in lesions in leaves, stems, and fruit. The lesions ooze bacterial cells that, when dispersed by windblown rain, can spread to other plants in the area. Infection may spread further by hurricanes. The disease can also be spread by contaminated equipment, and by transport of infected or apparently healthy plants. Due to latency of the disease, a plant may appear to be healthy, but actually be infected.

Citrus canker bacteria can enter through a plant's stomata or through wounds on leaves or other green parts. In most cases, younger leaves are considered to be the most susceptible. Also, damage caused by citrus leaf miner larvae (*Phyllocnistis citrella*) can be sites for infection to occur. Within a controlled laboratory setting, symptoms can appear in 14 days following inoculation into a susceptible host. In the field environment, the time for symptoms to appear and be clearly discernible from other foliar diseases varies; it may be on the order of several months after infection. Lower temperatures increase the latency of the disease. Citrus canker bacteria can stay viable in old lesions and other plant surfaces for several months.

Citrus canker lesions on fruit

## Detection

The disease can be detected in groves and on fruit by the appearance of lesions. Early detection is critical in quarantine situations. Bacteria can be tested for pathogenicity by inoculating multiple citrus species with them. Additional diagnostic tests (antibody detection), fatty-acid profiling, and genetic procedures using polymerase chain reaction can be conducted to confirm diagnosis and may help to identify the particular canker strain. Clara H. Hasse detected that citrus canker was not of fungoid origin but caused by bacteria. Her research published in the 1915 *Journal of Agricultural Research* played a major part in saving citrus crops in multiple states.

Susceptibility

Not all species and varieties of citrus have been tested for citrus canker. Most of the common species and varieties of citrus are susceptible to it. Some species are more susceptible than others, while a few species are resistant to infection.

| Susceptibility | Variety |
|---|---|
| Highly susceptible | Grapefruit (*Citrus* x *paradisi*), Key lime (*C. aurantiifolia*), Pointed leaf hystrix (*C. hystrix*), lemon (*C. limon*) |
| Susceptible | Limes (*C. latifolia*) including Tahiti lime, Palestine sweet lime; trifoliate orange (*Poncirus trifoliata*); citranges/citrumelos (*P. trifoliata* hybrids); tangerines, tangors, tangelos (*C. reticulata* hybrids); sweet oranges (*C. sinensis*); bitter oranges (*C. aurantium*) |
| Resistant | Citron (*C. medica*), Mandarins (*C. reticulata*) |
| Highly resistant | Calamondin (*X Citrofortunella*), kumquat (*Fortunella* spp.) |
| Modified from: Gottwald, T.R. et al. (2002). Citrus canker: The pathogen and its impact. Online. *Plant Health Progress* | |

## Management

Scientists have not been able to come up with a proper system to help treat outbreaks. If this disease continues to spread, farming citrus will become very costly and difficult.

## Distribution and Economic Impact

Citrus canker is thought to have originated in the area of Southeast Asia-India. It is now also present in Japan, South and Central Africa, the Middle East, Bangladesh, the Pacific Islands, some countries in South America, and Florida. Some areas of the world have eradicated citrus canker and others have ongoing eradication programs, but the disease remains endemic in most areas where it has appeared. Because of its rapid spread, high potential for damage, and impact on export sales and domestic trade, citrus canker is a significant threat to all citrus-growing regions.

## Australia

The citrus industry is the largest fresh-fruit exporting industry in Australia. Australia has had three outbreaks of citrus canker, all of which have been successfully eradicated. The disease was found twice during the 1900s in the Northern Territory and was eradicated each time. In 2004, an unexplained outbreak occurred in central Queensland. The state and federal governments ordered all commercial groves, all noncommercial citrus trees, and all native lime trees (*C. glauca*) in the vicinity of Emerald to be destroyed rather than trying to isolate infected trees. Eradication was successful, with permission to replant being granted to farmers by the biosecurity unit of the Queensland Department of Primary Industries in early 2009.

## Brazil

Citrus is an important domestic and export crop for Brazil. Citrus agriculture is the second-most important agricultural activity in the state of São Paulo, the largest sweet orange production area in the world. Over 100,000 groves are in São Paulo, and the area planted with citrus is increasing. Of the estimated 2 million trees, greater than 80% are a single variety of orange, and the remainder is made up of tangerine and lemon trees. Because of the uniformity in citrus variety the state has been adversely affected by canker, causing crop and monetary losses. In Brazil, rather than destroying entire groves to eradicate the disease, contaminated trees and trees within a 30-m radius are destroyed; by 1998, over half a million trees had been destroyed.

## United States

Citrus canker was first found in the United States in 1910 not far from the Georgia – Florida border. Subsequently, canker was discovered in 1912 in Dade County, more than 400 mi (600 km) away. Beyond Florida, the disease was discovered in the Gulf states and reached as far north as South Carolina. It took more than 20 years to eradicate that outbreak of citrus canker, from 1913 through 1931, $2.5 million in state and private funds were spent to control it—a sum equivalent to $28 million in 2000 dollars. In 26 counties, some 257,745 grove trees and 3,093,110 nursery trees were destroyed by burning. Citrus canker was detected again on the Gulf Coast of Florida in 1986 and declared eradicated in 1994.

The most recent outbreak of citrus canker was discovered in Miami, Dade County, Florida, on Sept. 28, 1995, by Louis Willio Francillon, a Florida Department of Agriculture agronomist. Despite eradication attempts, by late 2005, the disease had been detected in many places distant from the original discovery, for example, in Orange Park, 315 miles (500 km) away. In January 2000, the Florida Department of Agriculture adopted a policy of removing all infected trees and all citrus trees within a 1900-ft radius of an infected tree in both residential areas and commercial groves. Previous to this eradication policy, the department eradicated all citrus trees within 125 ft of an infected one. The program ended in January 2006 following a statement from the USDA that eradication was not feasible.

# Powdery Scab

Powdery scab is a disease of potato tubers. It is caused by the cercozoan *Spongospora subterranea* f. sp. *subterranea* and is widespread in potato growing countries. Symptoms of powdery scab include small lesions in the early stages of the disease, progressing to raised pustules containing a powdery mass. These can eventually rupture within the tuber periderm. The powdery pustules contain resting spores that release anisokont zoospores (asexual spore with two unequal length flagella) to infect the root hairs of potatoes or tomatoes. Powdery scab is a cosmetic defect on tubers, which can result in the rejection of these potatoes. Potatoes which have been infected can be peeled to remove the infected skin and the remaining inside of the potato can be cooked and eaten.

## Disease Cycle

In general, not a lot is known about the life cycle of *Spongospora subterranea* f.sp *subterranea* (Sss). Most of the currently-proposed life cycle is based on that of *Plasmodiophora brassicae*, a closely related and better-studied protozoan. It has been proposed, due to this similarity, that there are two distinct stages that Sss can exist as; the asexual and sexual stages.

Asexual Stage: A zoospore infects root tissue and becomes an uninucleate plasmodium. This plasmodium undergoes mitotic nuclear division (creates many nucli within a single cell) and turns into a multinucleate plasmodium. Then, the multinucleate plasmodium forms zoosporangium, which eventually release more zoospores. This process can happen relatively quickly and can act as an important source of secondary inoculum within a field.

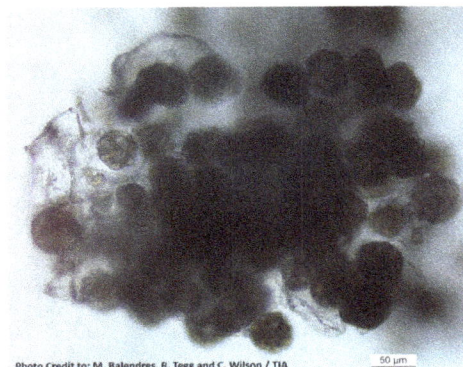

Photo Credit to: M. Balendres, R. Tegg and C. Wilson / TIA     50 µm

Sporosori (survival structure) of the powdery scab pathogen

Sexual Stage: This stage follows a similar pattern to the asexual stage, but with a few exceptions. It is hypothesized that two zoospores fuse together to form a dikaryotic zoospore (with two separate haploid nuclei, n+n) and then infect the roots. Once the infection occurs, the dikaryotic zoospore develops into a binucleate plasmodium (one pair on nuclei, n+n). Similar to the asexual stage, this plasmodium will also replicate its nucleus to create a multinucleate plasmodium (many pairs of nuclei, n+n). The second main different between stages occurs here. The pairs of nuclei (n+n) will fuse by karyogamy, and the plasmodium will quickly divide into numerous resting spores within a sporosori (spore sack, alternatively called cystosori). These resting spores have three-layered walls and are extremely resistant to the environment, allowing them to persist in the soil for longer than 10 years.

As a reminder, most of the life cycle is still unclear. However, the presence of zoospores, plasmodia, zoosporangia, and resting spores have been observed in the field and lab. The ploidy levels and karyogamy events are only theorized and have yet to be proven.

## Environment

One of the three factors of disease causation, as depicted in the disease triangle, is the environment. *Spongospora subterranea* pathogenesis is most effective in cool, damp environments, such as northern Britain, the Columbia Basin of south-central Washington, and north-central Oregon. The environmental condition is particularly critical during the release of infective agents (zoospores) into the soil-environment . Upon release from resting spores, zoospores require moisture to swim towards the host tuber or roots. One study, found powdery scab was significantly more common on plants grown in constant dampness compared to plants grown with varying moisture levels. In this same study she concluded disease risk was related more to the environment, or moisture level, than the level of inoculum present. Inoculum may be present but not able to disperse due to environmental conditions, and therefore does not reach host tissue to infect. Other environmental factors that affect *Spongospora subterranea* infection are directly related to agronomic practices. Increased use of fertilizers containing nitrate or ammonium nitrogen increase the incidence and severity of powdery scab. It is thought that the fertilization increases root growth, and thus provides more tissue for infection and disease cycling to occur. Also, reduced cellulose within the cell walls caused by excess nitrogen may increase susceptibility of host to infection. It is apparent that the environment can directly affect both the host susceptibility and the dispersal of the pathogen ultimately setting the pace for the disease cycle.

## Pathogenesis

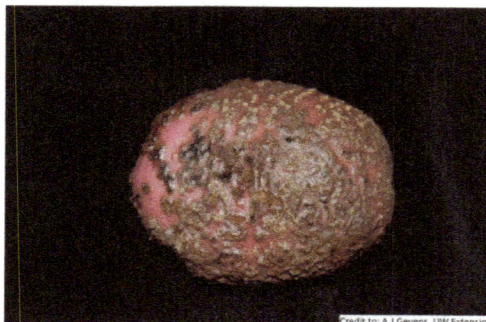

Potato tuber covered in powdery scabs

*S. subterranea* is an obligate parasite slime mold that infects the below ground structures of the host. Infection leads to hypertrophy and hyperplasia of the host cells and eventual bursting. However, the mechanism behind this is still unknown. Zoospores infect the root hairs by attaching to the outer surface, encysting, and then penetrating the epidermis through lenticels and stomata. Once inside, the multinucleate plasmodium divides to spread and produce more zoospores. The plasmodium causes the infected host cells to multiply rapidly and enlarge into a gall. This rapid multiplication also produces uninucleate cells that aggregate together as sporosori. The sporosori look like a powdery mass within the gall, which gives this disease its name. Eventually the gall swells and bursts out the epidermis of the tuber, releasing the spores back into the soil. Gall severity depends on inoculum level, environment, and potato skin type. Infection is most prevalent in the early stages of tuber formation while the potato tissue is unsuberized. But, infection can occur at all stages on development. White and red skinned potatoes and highly susceptible while russet skinned are somewhat resistant. Russet skin is thicker and has higher levels of the LOX protein which is used as a marker for resistance. There is little known about variation and sexual recombination within *S. subterranea*, therefore high priority is given to researching the variations within potato cultivars for researching host/pathogen relationships and management.

## Importance

Powdery Scab has important implications for commercial farming. Not only does the pathogen itself cause harm, but the pathogen is also a vector for potato mop-top virus, another plant pathogen. As a result, its presence greatly threatens potato yield for farmers. The burst pustules can also act as a wound for other fungi to infect, such as *Phytophthora erythroseptica* and *Phytophthora infestans*. Thus, tubers with powdery scab can have increased incidences of other devastating diseases, including pink rot, dry rot, black dot, and late blight. Potato tubers will form powdery scab pustules that inhibit their ability to be sold. Many markets decline to buy potatoes with ugly scarring even if they are safe to eat. Research has not yet found an effective way to peel the scabs without damaging the potato. Potatoes that are rejected for sale create a large financial burden on farmers. Additionally, because soil borne inoculum can survive for years as spores, the pathogen is very difficult to eliminate once present. In Great Britain a recent Potato Council funded diagnostic project discovered that as much as 82% of fields tested positive for soil inoculum.

## Management

*S. subterranea* currently has no effective chemical controls. Therefore, other cultural management techniques must be used. Using certified clean seeds and planting in fields that have been historically healthy is the best form of control. These methods may prevent infestation from resting spores. Since infection is promoted by cool soil temperatures and high soil moisture, delayed planting can also help reduce negative effects of the pathogen. Delayed planting reduces the growth period in cooler soils subsequently decreaseing germination of the spores. One limitation to this method is an additional decrease of early market yield. Pre-planting chemigation with metam sodium can reduce the propagules of the pathogen. Other common means of control include using resistant potatoes and crop rotations. Several cultivars of resistant potatoes include Granola, Nicola, Ditta, and Gladiator. Because soil-borne inoculum can survive for many years, crop rotations should involve alternate species that will promote a partial life cycle of the pathogen. This way the zoospores will germinate without producing new spores. Researchers have investigated the use of beta-aminobutyric acid (BABA) in promoting potato resistance. BABA triggers a plants systemic acquired resistance (SAR), a natural plant defense mechanism. When potatoes are inoculated with BABA and then later inoculated with the pathogen, *S. subterranean,* they exhibit overall reduction in disease. While pathogen reduction has been experimentally supported, further experimentation needs to be performed.

# Wilt Disease

Wilt diseaseA wilt disease is any number of diseases that affect the vascular system of plants. Attacks by fungi, bacteria, and nematodes can cause rapid killing of plants, large tree branches or even entire trees.

A pine tree with pine wilt

Wilt diseases in woody plants tend to fall into two major categories, those that start with the branches and those that start with the roots. Those that start with the branches most often start with pathogens that feed on the leaves or bark, those that start with the roots start with wounding or direct entry by the pathogen into the roots, some are spread from one plant to another by way of root grafts.

Pathogens that cause wilting diseases invade the vascular vessels and cause the xylem to fail to transport water to the foliage, thus causing wilting of stems and leaves.

## Wilt Diseases

Wilt diseases include:

## Bacterial Wilt of Cucurbits

Bacterial wilt of cucurbits is cause by the bacteria *Erwinia tracheiphila*, it affects cucumber, squash, muskmelon, pumpkin, gourds; certain varieties of cucumber and squash have different degrees of resistance. Once a plant is infected, the bacteria spread through the xylem vessels from the area of infection to the main stem, and the entire plant wilts and dies. Initial symptoms may include the wilting of single leaves and smaller stems. Infected plants may produce a creamy white bacterial ooze when cut. The bacteria survive winter in the digestive tract of striped cucumber beetles and spotted cucumber beetles. In the spring when the beetles are feeding on susceptible plants, the bacteria, which are contained in the fecal matter of the beetles enters the plant through wounds in the epidermis. The bacteria need a film of water to facilitate infection. The bacteria can also be transmitted from one plant to another when beetles feed on an infected plant and the bacteria becomes attached to the beetles mouthparts.

The bacteria *Ralstonia solanacearum* and related species cause bacterial wilt of bananas and plantains. The same bacteria also cause wilt diseases of potatoes (*Solanum tuberosum*), tomatoes (*Solanum lycopersicum*), aubergine (eggplant) (*Solanum melongena*), banana (*Musa* species), geranium (*Pelargonium* species), ginger (*Zingiber officinale*), tobacco (*Nicotiana tabacum*), sweet peppers (*Capsicum* species), olive (*Olea europea*), and others.

## Dutch Elm Disease

Dutch elm disease is caused by the fungus *Ophiostoma ulmi*, it affects elm trees.

## Elm Yellows

Elm yellows sometimes called elm phloem necrosis, affects elm trees and is caused by a Mycoplasma like organism. It is spread by the white-banded leafhopper.

## Mimosa Wilt

- caused by the fungus *Fusarium oxysporum*.

## Oak Wilt

Oak wilt is a fungal caused by *Ceratocystis fagacearum*, is a disease originating in eastern Russia. It can slowly or quickly kill an oak tree when the tree reacts to the fungus by plugging its own cambial tissue while attempting to block the spread of the fungus. This plug prevents the cambium vascular tissue from delivering nutrients and water to the rest of the plant, which eventually kills it. Red oaks are very susceptible.

## Persimmon Wilt

Persimmon wilt attacks persimmons and is caused by *Acromonium diospyri*, a fungus. In the United States it has nearly eliminated persimmons from the central basin of Tennessee. Because of its lethality to persimmons, it was proposed as a biological control agent to eliminate unwanted native persimmons.

## Pine Wilt

Pine wilt is caused by the North American native pinewood nematode (*Bursaphelenchus xylophilus*). Where it is indigenous it is not major pathogen of native pine species, but in North America it causes wilt in a few non-native North American pine species. It has been introduced into Japan and China, where it has become a troubling disease of Japanese red pines (*Pinus densiflora*) and black pines (*Pinus thunbergii*). Over 46 million cubic meters of trees have been lost alone in Japan over a 50-year period. It is spread among conifers by pine sawyer beetles (*Monochamus* spp). The nematodes can reproduce quickly in the sapwood under favorable conditions within susceptible pine species, causing wilting and death, sometimes in only a few weeks. North America lumber products are under export restrictions because of the nematode. In the Midwest United States it has killed many Scots pines (*Pinus sylvestris*), and this attractive tree is no longer recommended for landscaping uses there.

### Stewart's Wilt

Stewart's Wilt is caused by the bacteria *Pantoea stewartii* and affects corn plants especially sweet corn. It is a problem in the production of sweet corn in the Northeastern USA.

### Verticillium Wilt

Verticillium wilt affects over 300 species of eudicot plants caused by one of two species of Verticillium fungus, *V. dahliae* and *V. albo-atrum*. Many economically important plants are susceptible including cotton, tomatoes, potatoes, eggplants, peppers and ornamentals, as well as others in natural vegetation communities.

# Top Dying Disease

Top dying disease is a disease that affects *Heritiera fomes*, a species of mangrove tree known as "sundri", a characteristic tree of the estuarine complex of the Ganges–Brahmaputra Delta in Bangladesh and West Bengal. Although an increase in certain trace elements in the sediment deposited where these trees grow may be a factor in the incidence of the disease, its cause has not been fully established.

## History

About 63% of the trees growing in the Sundarbans are *Heritiera fomes*. Sporadic instances of top dying disease had been noticed in these trees earlier in the twentieth century but the disease became more acute and extensive after about 1970. An inventory of trees in the Sundarbans in 1985 recorded 45 million diseased trees with nearly half of these having more than half their crown affected. No causal agent has been discovered and the dieback is thought to be the result of stressful conditions, perhaps caused by an increase in the heavy metal concentration of the sediment deposited in the delta.

The maximum level of heavy metal contamination of sediments is found to take place in the late monsoon period and the concentration of these elements in the leaf litter varies with the season and does not necessarily correlate with the amount taken up by the trees. A 2014 study found that the successful establishment of seedlings of *H. fomes* and the rate of sapling growth were both adversely affected in areas with high levels of top dying disease, indicating that some environmental factors were likely to be significant. However, although there was some correlation between high levels of trace elements and the incidence of the disease, no particular elements could be demonstrated to cause the condition. The researchers concluded that heavy metal contamination might contribute to top dying disease, but was probably only one of several factors in its incidence.

## Symptoms

The uppermost parts of the tree are affected first with loss of leaves and dieback of branches in the crown. One or more knot-like swellings may develop on affected branches. Lower branches are progressively affected over time. Wood-boring insects and fungi invade the diseased wood and the tree eventually dies.

# Barley Yellow Dwarf

Barley yellow dwarf is a plant disease caused by the barley yellow dwarf virus, and is the most widely distributed viral disease of cereals. It affects the economically important crop species barley, oats, wheat, maize, triticale and rice.

## Biology

Barley yellow dwarf virus (BYDV) is a positive sense single-stranded RNA virus; the viron is not enveloped in a lipid coating. The virus is transmitted by aphids, and the taxonomy of the virus is based on genome organisation, serotype differences and on the primary aphid vector of each isolate.

The isolates and their major vectors (in parentheses) are:

- Subgroup I

    o MAV, a less severe strain carried by aphids (Grain aphid, *Sitobion avenae*), SGV (*Schizaphis graminum*), and PAV, a less severe strain carried by aphids (bird cherry-oat aphid, *Rhopalosiphum padi*, grain aphid, *S. avenae*, and others including rose-grain aphid, *Metopolophium dirhodum*).

- Subgroup II, called Cereal Yellow Dwarf Virus, however CYDV is now recognised as a separate species belonging to the *Polerovirus* genus of the Luteoviridae family

    o RPV, the most severe strain carried by aphids (Bird cherry-oat aphid, *Rhopalosiphum padi*), RMV (*Rhopalosiphum maidis*)

## Pathology

When aphids feed on the phloem of the leaf, the virus is transmitted to the phloem cells. Once inside the plant, the virus begins to replicate and assemble new virions. This process requires significant metabolic input from the plant, and causes the symptoms of barley yellow dwarf disease.

The symptoms of barley yellow dwarf vary with the affected crop cultivar, the age of the plant at the time of infection, the strain of the virus, and environmental conditions, and can be confused with other disease or physiological disorders. Symptoms appear approximately 14 days after infection. Affected plants show a yellowing or reddening of leaves (on oats and some wheats), stunting, an upright posture of thickened stiff leaves, reduced root growth, delayed (or no) heading, and a reduction in yield. The heads of affected plants tend to remain erect and become black and discoloured during ripening due to colonization by saprotrophic fungi. Young plants are the most susceptible.

Wheat plants dwarfed after infection with BYDV.

The virus is transmitted from the phloem when the aphid feeds. When the aphid feeds, virions go to the aphid's hind gut, the coat protein of the virus is recognised by the hindgut epithelium, and the virion is allowed to pass into the insect's hemolymph, where it can remain indefinitely, but the virus cannot reproduce inside the aphid. The virus is actively transported into the accessory salivary gland to be released into salivary canals and ducts. The virus is then excreted in the aphid saliva during its next feeding.

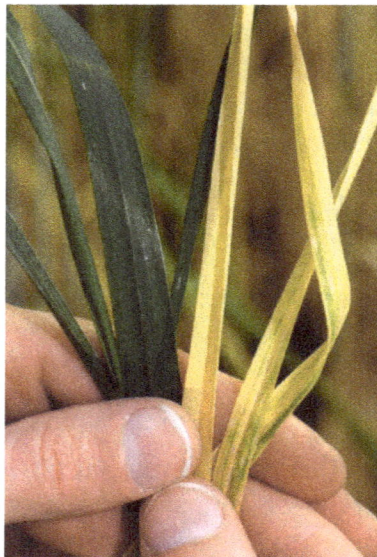

Infected wheat leaves have a reduced ability to photosynthesise.

The host range of BYDVs consists of more than 150 species in the Poaceae; a large number of grasses both annual and perennial are alternate hosts to BYVD and can serve as reservoirs of the virus.

## Sources and Spread

There are two main sources by which a cereal crop might be infected

1. By non-migrant wingless aphids already present in the field and which colonise newly-emerging crops. This is known as "green-bridge transfer".

2. By winged aphids migrating into crops from elsewhere. These then reproduce and the offspring spread to neighbouring plants.

## Effect On Yield

This is variable since it depends on viral strain, time of infection and rate of spread. Most severe losses are from early infections and can be as high as 50%.

## Control

"Green bridge" sources must be ploughed in as early as possible. Alternatively, a desiccant herbicide should be applied 10 days prior to cultivation. Insecticide sprays may be used at crop emergence.

Drilling dates prior to mid-October favors attacks from winged migrant aphids. However, yield penalties may be experienced from late drilling. Insecticide sprays in this instance are therefore aimed at killing the aphids before significant spread can occur.

## Products Used

Synthetic Pyrethroid Insecticides

# Raspberry Leaf Spot

Raspberry Leaf Spot is a plant disease caused by *Sphaerulina rubi*, an ascomycete fungus. Early symptoms of infection are dark green spots on young leaves. As the disease progresses, these spots turn tan or gray in color. Disease management strategies for Raspberry Leaf Spot include the use of genetically resistant raspberry plant varieties and chemical fungicide sprays.

Raspberries are an important fruit, mainly grown in Washington, Oregon and California. Although they are also grown in the Midwest and northeastern states, the output is not nearly as great due to the colder weathers and shorter growing seasons. *S. rubi* prefers warmer and wetter conditions, which can make raspberry production very difficult in California.

## Hosts and Symptoms

A Raspberry Leaf Spot infection initially causes dark green circular spots on the upper side of young leaves, which will eventually turn tan or gray. These spots are typically 1–2 mm in diameter, but can get as big as 4–6 mm. More severe infections can cause leaves to drop prematurely in the late summer and early fall. Due to the loss of leaves, infected raspberries are more susceptible to winter injury. As a result, Raspberry Leaf Spot may not only reduce yield in season, but cause lasting consequence into the next season.

The symptoms of Raspberry Leaf Spot are similar to the symptoms of Raspberry Anthracnose. The best way to differentiate between the two fungal diseases is to inspect the stems of the plant. Stem lesions are indicative of Raspberry Anthracnose.

In 1943, it was discovered that *S. rubi* only infects raspberry plants. Previously, the pathogen had also been blamed for leaf spot on blackberry and dewberry. However, Demaree and Wilcox demonstrated the raspberry pathogen could not cause leaf spots on blackberry or dewberry. The similar pathogens were also differentiated as perfect and imperfect, as the Blackberry Leaf Spot pathogen didn't have a known sexual stage.

## Disease Cycle

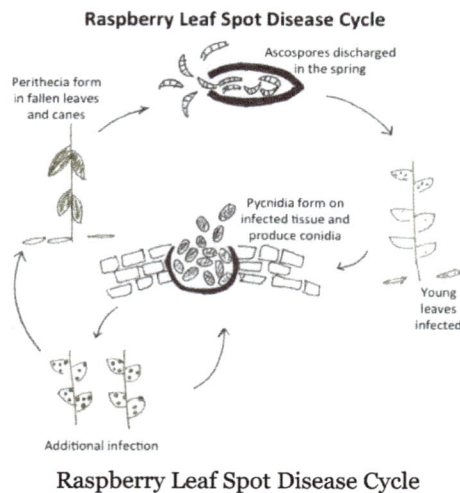

Raspberry Leaf Spot Disease Cycle

In spring, when conditions are favorable, ascospores are discharged from perithecia that have overwintered in fallen leaves and canes and disseminate to infect young leaves of raspberry plants. Once infected, the raspberry leaf serves as a nutrient source for the fungus to begin producing secondary inoculum, or conidia, within pycnidia, a survival structure that protects the spores. Conidia can undergo several repeating secondary cycles and re-infect other nearby plants. When the leaves of the raspberry plant begin to fall, perithecia form in the fallen tissue where asci and ascospores will be produced and protected until the following spring. The perithecia are black, found subepidermaly. The ascospores are characterized by a cylindrical, curved shape and are pointed at both ends with four septate usually.

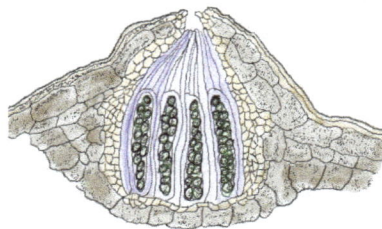

Diagram of a perithecium. Each ascus contains eight ascospores, shown in green.

## Environment

*S. rubi* grows optimally in humid conditions, which promotes wet leaves. In general, the conidia of *S. rubi* are disseminated through wind and rain. With these favorable conditions, the fungus can cause secondary infections more easily, thus leading to a more serious outbreak.

Furthermore, because the fungus produces pycnidia, a survival structure that contains conidia, it can survive in a range of temperatures, although the fungus grows optimally at 27 °C or 80 °F. Provided that there is adequate moisture, the conidia from the pycnidia will be able to disseminate via wind and rain.

Raspberry Leaf Spot infections will typically be more severe in parts of the United States that are climatically warmer and more humid.

## Management

Genetic resistance is the preferred disease management strategy because it allows farmers to minimize chemical intervention. Less pesticide and fungicide can encourage biological control agents, reduce production costs, and minimize the chemical residues in fruit. Some genetic varieties of raspberry are better than others for the control of leaf spot. Nova and Jewel Black are both resistant varieties, while Prelude and Honey Queen Golden Raspberry have some resistance, but can be susceptible depending on environmental conditions. Reiville, Canby, Encore and Anne are the most susceptible varieties.

Cultural practices are also important for the management of Raspberry Leaf Spot. Sanitation, which includes the removal of all plant debris and infected canes in the fall, reduces places for the pathogen to overwinter. Pruning the raspberry plants and planting in rows will allow for airflow to dry leaves, creating an uninviting environment for fungi. Furthermore, air flow circulation is important for reducing sporulation and successful infection. Lastly, avoid wounding the plants, as this may provide the fungus with an opportunity to infect.

## Importance

Raspberry is the third most popular berry in the United States. In the US, per capita consumption of fresh raspberries was 0.27 pounds in 2008 with frozen raspberry consumption adding 0.36 pounds. Not only are raspberries consumed as a snack, but they have historically also been used for much more. Parts of the raspberry plant used to be used to treat several ailments, including as a relief for painful menstrual cramps, an aid in childbirth, a cure for diarrhea, a remedy for heart disease, and even as a preventative measure for vomiting.

Although there is a high demand for raspberries, growers find it very difficult to grow them. Not only are they relatively fastidious when it comes to general requirements for survival, but they also tend to be susceptible to disease. Specifically, Raspberry Leaf Spot can be a debilitating disease if conditions are favorable. If defoliation does occur due to Raspberry Leaf Spot, the outcome can be economically devastating for the farmer. Defoliation would cause the loss of the plant's ability to photosynthesize, and thus, the fruit would be lost shortly after. Yield for raspberries can be anywhere from 0 to 6,000 pounds/acre, typical yields being 4,000 to 5,000 pounds/acre. With an input cost of approximately $4,000, raspberries are a risky endeavor.

# References

- Bhatt, J.C. (1998). "Plant Diseases and Their Management". In Pande, D.C. Managing Agriculture for a Better Tomorrow: The Indian Experience. M.D. Publications Pvt. Ltd. p. 62. ISBN 9788175330672.

- Cranshaw, W. (1998). Pests of the West, 2nd Edition: Prevention and Control for Today's Garden and Small Farm. Fulcrum Publishing. p. 181. ISBN 9781555914011.

- Clay, H.F.; Hubbard, J.C. (1987). Tropical Shrubs. The Hawai'i Garden Volume 2. University of Hawaii Press. p. 30. ISBN 9780824811280.

- Foucher, F. (2009). "Functional Genomics in Rose". In Folta, K.M.; Gardiner, S.E. Genetics and Genomics of Rosaceae. Springer. p. 387. ISBN 9780387774909

- Wale, Stuart; Platt, Bud; Cattlin, Nigel D. (2008-04-11). Diseases, Pests and Disorders of Potatoes: A Colour Handbook. CRC Press. ISBN 9781840765083.

- Prior, Philippe; Allen, Caitilyn; Elphinstone, John (1998). Bacterial Wilt Disease : Molecular and Ecological Aspects. Springer. p. 6. ISBN 978-3-540-63887-2.

- "Bulletin #2436, Powdery Scab of Potatoes | Cooperative Extension Publications | University of Maine". umaine.edu. Retrieved 2015-11-11.

- Gábor, Z. "Form as Movement in Goethe's The Metamorphosis of Plants". Technical University of Budapest. Retrieved 3 November 2012.

# Various Diseases of Wheat and Leaf

The chapter explicates diseases such as glomerella graminicola, common bunt, spot blotch, wheat leaf rust, and leaf rust. Spot blotch is a leaf disease of wheat caused by Cochliobolus sativus while leaf rust is a fungal disease of barley caused by Puccinia hordei. Also known as brown rust, it is the most serious rust disease on barley. The aspects elucidated in this chapter are of vital importance, and provide a better understanding of plant pathology.

## Glomerella Graminicola

Glomerella graminicola ((anamorphic) *Colletotrichum graminicola*) is a fungus in the teleomorphic phase whose anamorphic phase, *Colletotrichum graminicola,* causes anthracnose in many cereal species including maize and wheat. Corn is affected in large numbers in the United States by this fungus, especially certain varieties that have been genetically engineered. These engineered varieties are more susceptible to the teleomorph phase of the fungus. It is not until the fungus moves to the teleomorph phase of the lifecycle and begins to produce fruiting bodies that host plants will begin to exhibit symptoms, often on plants depleted in energy after the stress of pollination. Once the pathogen is in a field, producers can suffer huge economic losses. The disease, corn anthracnose leaf blight, is the most common stalk disease in maize and occurs most frequently in reduced-till or no-till fields. As these practices are widespread, as can be the pathogen.

Anthracnose stalk rot

## Host and Symptoms

*C. graminicola* is a fungal pathogen that colonizes and infects many turfgrass e.g., Bluegrass, Rye-grass, fescue. In addition to the grasses, C. *graminicola* also infects many grain crops such as barley, wheat, sorghum and corn. The fungus can infect many different parts of the corn plant, typically the kernels, tassels, roots, leaves, stalk and husks. The most common area of infection is the stalk. C. *graminicola* produces three major symptom types: leaf blight, stalk rot and top die-back. The leaf blight is characterized by round yellowing water soaked lesions on the leaves. These lesions usually occur early in the season and are how this pathogen is distinguished from other diseases. Top die-back is the necrosis of the top leaves and stalk of the corn. This occurs around the same time as grain formation. The stalk rot phase becomes prominent during the late reproductive stages of the corn life cycle. It is characterized by blackening of the pith tissue in the stalk and also of the rind, beginning at the nodes closest to the soil. Along with these symptoms, seedling blight and post emergence damping off are also found.

## Identification

### Stromata

- 70-300 μm in diameter

- Bear prominent, dark, septate spines (setae) up to 100 μm long.

### Conidia

- Developing at the base of the spines

- Hyaline to pale yellow, unicellular, sickle-shaped, falcate to fusiform, tapered toward both ends

- 3-5 x 19-29 μm.

### Phialides

- Unicellular, hylanine and cylindrical,

- 4-8 x 8-20 μm.

### Growth on PDA

- Gray and feltlike

- Conidia and appressoria are numerous when culture are well aerated, and sclerotia sometimes occur.

- Appressoria are diagnostic: they are tawny brown, irregular-shaped in edge, prominent, and terminal on thickened hyphae.

## Disease Cycle

In the spring, fruiting structures (acervuli) form from corn residue and produce spores (conidia) that are dispersed by wind blown raindrops and splashing. Conidial spores infect young plants through the epidermis or stomata. Anthracnose develops rapidly in cloudy, overcast conditions with high temperatures and humidity. In optimal environmental conditions, conidia can germinate in as little as 6–8 hours in 100% humidity. Initial necrotic spots or lesions can be seen within 72 hours after infection by conidia. Lower leaves that develop lesions provide conidial spores and cause secondary infections on the upper leaves and stalk. Vascular infections primarily occur from wounds caused by stalk-boring insects, such as the larvae of the European corn borer, allowing for conidia to infect and colonize the xylem. From this, anthracnose top die back (vascular wilt) or stalk rot can occur. In the fall, *C. graminicola* survives as a saprophyte on corn leaf residue. The pathogen can also overwinter on corn stalks as conidia in an extracellular secretion. The secretion prevents conidia from desiccating and protects them from unfavorable environmental conditions. Overwintering on corn residue serves as a vital source of primary inoculum for the leaf blight phase in the spring. The cycle will start all over again when susceptible corn seedlings emerge from the ground in the spring.

## Disease Management

Since C. graminicola is found to survive on corn residue, specifically on the soil surface; one of the most effective methods of control is a one-year minimum of crop rotation to reduce anthracnose leaf blight. A study in 2009 showed more severe symptoms of leaf blight due to C. graminicola when grown on fields previously used for corn in comparison to fields previously used for soybean. Other management methods include the use of hybrid selections and tillage systems. Keeping in mind, hybrid selection may be resistant to leaf blight but they are not necessary resistant to other fungal diseases such as stalk rot. Tillage systems that are able to fully bury corn residue deep underground along with one year crop rotation will reduce the source of inoculum greatly. More work is still needed in order to determine the influence of buried and surface corn residues as a source of inoculum for corn anthracnose.

## Importance

Corn anthracnose caused by *Colletotrichum graminicola* is a disease present worldwide. This disease can affect all parts of the plant and can develop at any time during the growing season. This disease is typically seen in leaf blight or stalk rot form. Before the 1970s, Anthracnose was not an issue in North America. In the early 1970s, north-central and eastern U.S was hit with severe epidemics. Within 2 years of *C.graminicola's* appearance in Western-Indiana sweet corn production for canning companies were nearly wiped out and production no longer exists there today.

Anthracnose stalk rot was seen in many U.S corn fields in the 1980s and 1990s. A survey conducted in Illinois in 1982 and 1983 found that 34 to 46% of rotted corn stalks contained *C. graminicola*. Estimates on yield grain losses from anthracnose leaf blight and stalk rot range from zero to over 40%. This is dependent on hybrid, environment, timing of infection, and other stresses.

The impacts of *C. graminicola* are predicted to increase as the use of Bt corn becomes more common. Bt engineered corn has been seen to have a greater proportion of stalk rot and be more susceptible to *C. graminicola* compared to strains without Bt.

## Pathogenesis

Once conidia germinate on corn leaves, a germ tube differentiates and develops into an appresoria and allows *C. graminicola* to penetrate epidermal cells. Germination and appressorium formation occur best in the temperature range (15-30°C) Interestingly, penetration occurs in a much narrower temperature range (25-30°C). In order to penetrate the cell wall, the fungus first pumps melanin into the walls of the appressorium to create turgor pressure in the appressorium. The melanin allows water into the appressorium cell but nothing out. This builds up an incredible amount of turgor pressure which the fungus then uses to push a hyphae through the corn cell wall. This is called the penetration peg. The penetration peg then grows, extends through the cell extracting nutrients and the host cell wall dies. Hyphae migrate from epidermal cells to mesophyll cells. As a defense response, the cells produce papillae to prevent cell entry but is typically not seen successful. It is believed C. graminicola has a biotrophic phase because the plasma membrane of the epidermal cells is not immediately penetrated after invasion into the epidermal cell wall. Between 48–72 hours after infection, *C. graminicola* shifted from biotrophic growth to nectrotrophy (lesions appear). This is when secondary hyphae invade cell walls and intercellular spaces.

## Common Bunt

Common bunt, also known as stinking smut and covered smut, is a disease of both spring and winter wheats. It is caused by two very closely related fungi, *Tilletia tritici* (syn. *Tilletia caries*) and *T. laevis* (syn. *T. foetida*).

## Symptoms

Plants with common bunt may be moderately stunted but infected plants cannot be easily recognized until near maturity and even then it is seldom conspicuous. After initial infection, the entire kernel is converted into a sorus consisting of a dark brown to black mass of teliospores covered by a modified periderm, which is thin and papery. The sorus is light to dark brown and is called a bunt ball. The bunt balls resemble wheat kernels but tend to be more spherical. The bunted heads are slender, bluish-green and may stay greener longer than healthy heads. The bunt balls change to a dull gray-brown at maturity, at which they become conspicuous. The fragile covering of the bunt balls are ruptured at harvest, producing clouds of spores. The spores have a fishy odor. Intact sori can also be found among harvested grain.

## Disease Cycle

Millions of spores are released at harvest and contaminate healthy kernels or land on other plant parts or the soil. The spores persist on the contaminated kernels or in the soil. The disease is initiated when soil-borne, or in particular seed-borne, teliospores germinate in response to moisture and produce hyphae that infect germinating seeds by penetrating the coleoptile before plants emerge. Cool soil temperatures (5° to 10°C) favor infection. The intercellular hyphae become established in the apical meristem and are maintained systemically within the plant. After initial infection, hyphae are sparse in plants. The fungus proliferates in the spikes when ovaries begin to form. Sporulation occurs in endosperm tissue until the entire kernel is converted into a sorus consisting of a dark brown to black mass of teliospores covered by a modified periderm, which is thin and papery.

## Pathotypes

Well-defined pathogenic races have been found among the bunt population, and the classic gene-for-gene relationship is present between the fungus and host.

## Management

Control of common bunt includes using clean seed, seed treatments chemicals and resistant cultivars. Historically, seed treatment with organomercury fungicides reduced common bunt to manageable levels. Systemic seed treatment fungicides include carboxin, difenoconazole, triadimenol and others and are highly effective. However,in Australia and Greece, strains of *T. laevis* have developed resistance to polychlorobenzene fungicides.

# Spot Blotch (Wheat)

Spot blotch is a leaf disease of wheat caused by *Cochliobolus sativus*. *Cochliobolus sativus* also infects other plant parts and in conjunction with other pathogens causes common root rot and black point.

## Introduction

Foliar blight or Helminthosporium leaf blight (HLB) or foliar blight has been a major disease of wheat (Triticum aestivum L.) worldwide. Foliar blight disease complex consists of spot blotch and tan spot. Spot blotch is favored in warmer environments whereas tan spot is favored in cooler environments such as United States. The tan spot forms of foliar blight appears in United States causing significant yield loss. With changed climatic conditions the disease is supposed to be increasing in cooler parts of the world. Among foliar blights the tan spot, caused by Pyrenophora tritici-repentis, is the most destructive leaf spot disease found in all wheat classes throughout the growing season across North Dakota.

The spot blotch form of foliar blight is severe particularly in warmer growing areas characterized by an average temperature in the coolest month above 17 °C. In the past 20 years, HLB has been recognized as the major disease constraint to wheat cultivation in the warmer eastern plains of South Asia. 25 millions of non-traditional wheat growing area are under the pressure of the disease.

## Symptoms

Early lesions are characterized by small, dark brown lesions 1 to 2 mm long without chlorotic margin. In susceptible genotypes, these lesions extend very quickly in oval to elongated blotches, light brown to dark brown in colour. They may reach several centimetres before coalescing and inducing the death of the leaf. Fruiting structures develop readily under humid conditions and are generally easily observed on old lesions. If spikelets are affected, it can result in shrivelled grain and black point, a dark staining of the embryo end of the seed. The small dark brown spots on the leaves contrast with the larger, light brown spots or blotches produced by tan spot and septoria avenae blotch.

## Typical Spot Blotch Sysmtom

Spot blotch symptom

Tan spot pathogen

Foliar blight associated pathogens

## Crop Losses

In recent years, Helminthosporium leaf blights (HLB), caused by both *Cochliobolus sativus* and *Pyrenophora tritici-repentis*, have emerged as serious concerns for wheat cultivation in the developing world. The disease causes significant yield losses overall 22% to complete failure of crop under severe epidemics.

## Control Measures

The disease is very serious in different parts of the world. The management of this disease requires integrative approach.

## An Integrated Approach

The best way to control Helminthosporium diseases is through an integrated approach. It includes the use of a variety of resistance sources, such as hexaploid wheat from Brazil and China (some of which is rate-limiting), alien genes and synthetic wheats. In addition, appropriate management practices that enhance the health of the plant populations, in general, are critical. Cooperation of pathologists, breeders and agronomists will be necessary to ensure sustainable control of this group of diseases. Economic feasibility of recommended practices has to be determined as part of the research. Options for controlling tan spot and spot blotch include disease-free seed, seed treatment with fungicides, proper crop rotation and fertilization, cultural practices in order to reduce inoculum sources, the use of chemicals and the research of disease resistance. The latter offers the best long-term control at no cost for the farmer and is ecologically safe.

## Seed Health

In Brazil, it is recommended not to plant seed lots with more than 30 percent black point in order to limit spot blotch. Seed treatment may prove to be appropriate, although the inoculum remaining on secondary hosts or in the soil may reduce the treatment efficiency. Seed treatments with phytoalexin inducer appeared to provide good protection to wheat seedlings against '**B. sorokiniana**' infection (Hait and Sinha, 1986). Seed treatment with fungicide will help protect germinating seed and seedlings from fungi causing seedling blights. Fungicide seed treatments include: captan, mancozeb, maneb, thiram, pentachloronitrobenzene (PCNB) or carboxin guazatine plus, iprodione and triadimefon (Stack and McMullen, 1988; Mehta, 1993). Seed-borne inoculum of '**P.**

*tritici-repentis'* can be controlled with seed-applied fungicides, such as guazatine and guazatine + imazalil, but other chemicals are also effective (Schilder and Bergstrom, 1993).

## Rotations and Crop Management

Clearing or ploughing in the stubble, grass weeds and volunteer cereals reduce inoculum as does crop rotation (Diehl et al., 1982). Reis et al. (1998) indicate that eradicant fungicide treatment of the seed and crop rotation with non-host crops can control spot blotch. In the rice-wheat system of South Asia, little work has been done on the epidemiology of HLB and how management of the rotation crops affects spot blotch and tan spot, except as noted earlier. More quantitative information is required on the role of alternate rotations, soil and plant nutrition, inoculum sources and climate. In the rice-wheat system, there is a need for timely planting of wheat, better stand establishment and root development, increased soil organic matter, sufficient levels of macro- and micronutrients, and water and weed management (Hobbs et al., 1996; Hobbs and Giri, 1997). Crop rotation and organic manures will play a major role in HLB. This should favour beneficial soil organisms as well as better plant nutrition. In the rice-wheat system, it will be necessary to break the rotation with other crops to make it more sustainable, and this should help reduce disease problems in general. The use of oilseed rape in South Asia is common in mixture with wheat or in rotation. Since rape is known to have some fungitoxic effects upon decay, its effects on HLB would need research (Dubin and Duveiller, 2000). In the HLB complex, rotations would need to be sufficiently long to reduce the amount of soil inoculum. Cook and Veseth (1991) note that the kind of rotation crop may not be so important to root health as the length of time out of wheat. The rotation crops and length of rotation would have to be studied in relation to HLB.

Apparently, sound management recommendations may antagonize specific diseases as in the case of tan spot. Tan spot has been controlled largely by cultural practices, such as rotation with non-host crops and removal or burial of stubble (Rees and Platz, 1992). Bockus and Claassen (1992) observed that rotation to sorghum was as effective as ploughing for control of tan spot, and under certain conditions, crop rotations as short as one year controlled tan spot. In South Asia, recent work by Hobbs and Giri (1997) indicates that minimum tillage may be the best way to reduce turnaround time from rice to wheat and thus permit the planting of wheat more timely. Since this probably increases inoculum of tan spot, it highlights the need for integration of disciplines to determine how best to achieve attainable yields.

## Fungicides

Although pesticide use should be minimized, fungicides have proven useful and economical in the control of tan spot (Loughman et al., 1998) and spot blotch (Viedma and Kohli, 1998). The triazole group (e.g. tebuconazole and propiconazole) especially has proven to be very effective for both HLBs, and their judicious use should not be overlooked. However, it may provide acceptable control but not always economic return in commercial grain production. This is dependent on the price received for the wheat, the price of the fungicide and the percent yield increase from using the fungicide. Situations will differ significantly according to geographical areas and cropping conditions. Spot blotch in particular is a very aggressive disease, and under a favourable environment, spraying at one- to two-week intervals for as long as necessary may be needed to maintain the disease under control.

For general information on management of the disease visit Ohio State University Link and FAO link

## Breeding For Resistance

The wheat cultivars of South Asia have only low to moderate levels of resistance to spot blotch. However, genetic variation for resistance has been reported in a few wheat. The best sources of resistance, to date, were identified in the Brazilian and Zambian wheat lines. Recently, a few Chinese wheat genotypes from the Yangtze river valley were identified with acceptable levels of resistance to spot blotch. The following genotypes has been reported to have satisfactory level of resistance, although complete resistance or immunity is lacking

1 SW 89-5193

2 SW 89-3060

3 SW 89-5422

4 Chirya 7

5 Ning 8319

6 NL 781

7 Croc 1/A. sq.// Borl

8 Chirya 3

9 G162

10 Chirya 1

11 Yangmai-6

12 NL 785

The field resistance governed by Chirya-3 and Milan / Sanghai 7 was found under monogenic control

Similarly resistant genotypes Acc. No. 8226, Mon/Ald, Suzhoe#8 from India are found to possess three genes for resistance.

A study was conducted to determine microsatellite markers associated with resistance in the F7 progeny from a cross between the spot blotch-susceptible Sonalika and resistant G162 wheat genotypes. 15 polymorphic markers showed association with two bulks, one each of progeny with low and with high spot blotch severity.

One of interesting phenomena associated with foliar blight in some of susceptible cultivar is tolerance (low yield loss even at very high level of disease severity). In addition, the resistance seems to be associated with late maturity (which is undesirable character as late maturing genotypes need to face more heat stress than early ones), complete understanding of physiological association may aid to complete understanding of the host-pathogen system.

CIMMYT wheat pathologist Dr. Duveiller and Rosyara at a spot blotch screening nursery at Rampur

Rosyara et al. reported that the AUDPC showed a significant negative correlation with the width of large vascular bundles, percentage of small vascular bundles with two girders and the number of large veins. Also the AUDPC was positively correlated with the distance between adjacent vascular bundles and leaf thickness. The chlorophyll or general health indicators, SPAD and AUSDC values were higher in spot blotch resistant and tolerant genotypes. The findings the study underlined the importance of mesophyll structure and chlorophyllcontent in spot blotch resistance in wheat. Also tolerant genotypes responded in the same way as artificial defoliation showing mechanisms of nutrient balance playing role. Similarly, canopy temperature depression was found associated with foliar blight resistance. Leaf tip necrosis was found to be associated with foliar blight resistance and is suggested as phenotypic marker. Different studies are done to estimate heritability and increase selection efficiency. Heritability estimates were low to high in terms of AUDPC. To increase efficiency of selection use of selection index has been suggested. The index includes days to heading (maturity related trait), thousand kernel weight, and area under foliar blight disease progress curve.

## Wheat Leaf Rust

Wheat leaf rust is a fungal disease that affects wheat, barley and rye stems, leaves and grains. In temperate zones it is destructive on winter wheat because the pathogen overwinters. Infections can lead up to 20% yield loss exacerbated by dying leaves which fertilize the fungus. The pathogen is Puccinia rust fungus. *Puccinia triticina* causes 'black rust', *P.recondita* causes 'brown rust' and *P.striiformis* causes 'Yellow rust'. It is the most prevalent of all the wheat rust diseases, occurring in most wheat growing regions. It causes serious epidemics in North America, Mexico and South America and is a devastating seasonal disease in India. All three types of *Puccinia* are heteroecious requiring two distinct and distantly related hosts (alternate hosts). Rust and the similar smut are members of the class Pucciniomycetes but rust is not normally a black powdery mass.

## Host Resistance

Plant breeders have tried to improve yield quantities in crops like wheat from the earliest times. In recent years, breeding for the resistance against disease proved to be as important for total wheat production as breeding for increase in yield. The use of a single resistance gene against various pests and diseases plays a major role in resistance breeding for cultivated crops. The earliest single resistance gene was identified as effective against yellow rust. Numerous single genes for leaf rust resistance have since been identified, the 47th genes prevent crop losses due to *Puccinia recondite* Rob. Ex Desm. f.sp. *tritici* infections, which can range from 5–15% depending on the stage of crop development.

Leaf rust resistance gene is an effective adult-plant resistance gene that increases resistance of plants against *P. recondita* f.sp. *tritici* (UVPrt2 or UVPrt13) infections, especially when combined with genes Lr13 and gene Lr34 (Kloppers & Pretorius, 1997). Lr37 originates from the French cultivar VPM1 (Dyck & Lukow, 1988). The line RL6081, developed in Canada for Lr37 resistance, showed seedling and adult-plant resistance to Leaf, yellow and stem rust. Crosses between the French cultivars will therefore introduce this gene into local germplasm. Not only will the gene be introduced, but the genetic variation of South African cultivars will also increase.

Molecular techniques have been used to estimate genetic distances among different wheat cultivars. With the genetic distances known predictions can be made for the best combinations concerning the two foreign genotypes carrying gene Lr37, VPMI and RL6081 and local South African cultivars. This is especially important in wheat with its low genetic variation. The gene will also be transferred with the least amount of backcrosses to cultivars genetically closest to each other, generation similar genetic offspring to the recurrent parent, but with gene Lr37, Genetic distances between near isogenic lines (NILs) for a particular gene will also give an indication of how many loci, amplified with molecular techniques, need to be compared in order to locate putative markers linked to the gene.

## Nomenclatural History

What is the appropriate name for Wheat Leaf Rust?

Fungal names are important. These are the keys to all information behind them. Then, an appropriate name can lead users to the right information. In the case of plant pathogenic fungi using an appropriate name is more important because of practical reasons. There are several examples among rust fungi of one species called with different names during different eras. However, one of the most interesting ones is the name for *Puccinia* species causing Wheat Leaf Rust (WLR). This species has been called by at least six different names since 1882, when G. Winter (1882) described the *Puccinia rubigo-vera*. For long time WLR interpreted as a specialized form of *P. rubigo-vera*. Later, Eriksoon and Henning (1894) put it under the *P. dispersa* f.sp. *tritici*. In 1899 and after some experiments Eriksson concluded that the rust should be considered as a separate authentic species. For this reason he described *P. triticina*. This name was used by Gaeumann (1959) in his comprehensive book about rust fungi of middle Europe. Mains (1933) was among the first scientists who used a species name with broad species concept for WLR. He considered *P. rubigo-vera* as current name and put 32 binomials as synonyms of that species. The next important article about naming WLR was published by Cummins and Caldwell (1956). They considered the same broad species concept and also discussed the validity of *P. rubigo-vera* which was based on an uerdinial stage basionym. Finally, they introduced P. recondita as the oldest valid name for WLR and also other grasses. Their idea and publication was followed by Wilson & Henderson (1966) in another comprehensive rust flora viz. British Rust Flora. Wilson and Henderson (1966) also used a broad species concept for *P. recondita* and divided this broad species to 11 different formae speciales. The accepted name for WLR in their flora was *P. recondita* f.sp. *tritici*.

Cummins (1971) in his rust monograph for Poaceae introduced an ultra-broad species concept for *P. recondita* and listed 52 binomials as its synonyms. Such a concept found great attention among mycologists and plant pathologists around the world and that is the reason we still can see *P. recondita* as an appropriate name for WLR in some publications. There was another stream opposite to broad morphologically-based concept among uredinologists. In the case of graminicolous rust fungi this stream was started by Urban (1969) who introduced *P. perplexans* var. *triticina* as an appropriate name for WLR. To Urban's understanding, a taxonomic name should reflect both morphology and ecology of the species. Savile (1984) was also among the uredinologists believing in narrowing the species concept and considered *P. triticina* as an authentic taxonomic name for WLR. Urban's research continued and he put many morphological, ecological and also field experiences together. Finally he considered WLR as a part of Puccinia persistens species with aecial stage on Ranunculaceae members, totally different from *P. recondita* which produces its aecial stage on Boraginacec family members. His final name for this rust was *P. persistens* subsp. *triticina*. Interestingly, recent molecular and also morphological studies proved Urban's taxonomy for WLR. It seems after more than a century and after introducing several names, we have an appropriate name for WLR.

## Life Cycle

Wheat leaf rust spreads via airborne spores. Five types of spores are formed in the life cycle. Uredospores, teleutospores, and basidiospores develop on wheat plants and pycnidiospores and aeciospores develop on the alternate hosts. The germination process requires moisture, and works best at 100% humidity. Optimum temperature for germination is between 15–20 °C. Before sporulation, wheat plants appear completely asymptomatic. In the Asian Subcontinent, the spores cannot survive the hot, dry weather but are re-introduced every year from the Himalayas or surrounding hills, possibly coming from *Berberis spp*, *Thalictrum flavum* and *Muehlenbergia huglet* which is a main reason for bread mouldes or even some grasses. Wheat rust pathogens are biotrophic and require living plant cells to survive.

*P. triticina* has an asexual and sexual life cycle. In order to complete its sexual life cycle *P. triticina* requires a second host *Thalictrum spp.* on which it will overwinter. In places where *Thalictrum* does not grow, such as Australia, the pathogen will only undergo its asexual life cycle and will overwinter as mycelium or uredinia. The germination process requires moisture and temperatures between 15–20 °C. After around 10–14 days of infection, the fungi will begin to sporulate and the symptoms will become visible on the wheat leaves.

The pathogen has an asexual and sexual cycle. In North America, South America and Australia the pathogen only undergoes its asexual cycle. However this does not seem to be a disadvantage to it, and wheat leaf rust has many races with different virulence. The sexual life cycle of wheat leaf rust requires a different host species, *Thalictrumn spp.*

## Symptoms

Small brown pustules develop on the leaf blades in a random scatter distribution. They may group into patches in serious cases. Infectious spores are transmitted via the soil. Onset of the disease is slow but accelerated in temperatures above 15 °C, making it a disease of the mature cereal plant in summer, usually too late to cause significant damage in temperate areas. Losses of between 5 and 20% are normal but may reach 50% in severe cases.

## Control

Varietal resistance is important. Chemical control with triazole fungicides may be useful for control of infections up to ear emergence but is difficult to justify economically in attacks after this stage

# Bacterial Leaf Scorch

Bacterial leaf scorch (commonly abbreviated BLS, also called bacterial leaf spot) is a disease state affecting many crops, caused mainly by the xylem-plugging bacterium *Xylella fastidiosa*. It can be mistaken for ordinary *leaf scorch* caused by cultural practices such as over-fertilization.

## Hosts

BLS can be found on a wide variety of hosts, ranging from ornamental trees (elm, maple, oak) and shrubs, to crop species including blueberry and almond.

## Bacterial Spot of Peppers and Tomatoes

Bacterial spot of peppers and tomatoes is caused by the bacteria *Xanthomonas campestris pv. Vesicatoria*.

## Bacterial Spot of Peaches

Bacterial spot of peaches is caused by the bacteria *Xanthomonas campestris pv. Pruni*. Spots may form on the leaves and they can be mistaken for *peach scab*, which is caused by a fungus.

## Symptoms

An irregular browning leaf margin which may or may not be bordered by a pale halo.

Symptoms re-occur every year, spreading throughout the tree crown, eventually killing the host plant.

## Vectors

Xylem-feeding leafhoppers can transmit the disease bacteria

## Treatment

There are no known effective treatments for BLS, consequently, removal of affected plants is recommended

# Black Sigatoka

Black sigatoka is a leaf-spot disease of banana plants caused by the ascomycete fungus *Mycosphaerella fijiensis* (Morelet). Also known as black leaf streak, it was discovered in 1963 and named for its similarities with the yellow sigatoka, which is caused by *Mycosphaerella musicola* (Mulder), which was itself named after the Sigatoka Valley in Fiji, where an outbreak of this disease reached epidemic proportions from 1912 to 1923. Plants with leaves damaged by the disease may have up to 50% lower yield of fruit and control can take up to 50 sprays a year.

## Life History

*M. fijiensis* reproduces both sexually and asexually, and both conidia and ascospores are important in its dispersal. The conidia are mainly waterborne for short distances, while ascospores are carried by wind to more remote places (the distances being limited by their susceptibility to ultraviolet light). Over 60 distinct strains with different pathogenetic potentials have been isolated. To better understand the mechanisms of its variability, projects to understand the genetic diversity of *M. fijiensis* have been initiated.

When spores of *M. fijiensis* are deposited on a susceptible banana leaf, they germinate within three hours if the humidity is high or a film of water is present. The optimal temperature for germination of the conidia is 27 °C (81 °F). The germ tube grows epiphytically over the epidermis for two to three days before penetrating the leaf by a stoma. Once inside the leaf, the invasive hypha forms a vesicle and fine hyphae grow through the mesophyll layers into an air chamber. More hyphae then grow into the palisade tissue and continue on into other air chambers, eventually emerging through stomata in the streak that has developed. Further epiphytic growth occurs before the re-entry of the hypha into the leaf through another stoma repeats the process. The optimal conditions for *M. fijiensis* as compared with *M. musicola* are a higher temperatures and higher relative humidity, and the whole disease cycle is much faster in *M. fijiensis*.

## Symptoms

Most infections start on the underside of the leaf. The symptoms start as small specks that become streaks running parallel to the leaf veins. These streaks aggregate and eventually form spots that coalesce, form a chlorotic halo, and eventually merge to cause extensive necrosis.

## Commercial Effect

The world-wide spread of the disease has been rapid, with its naming and first reported occurrence in 1963. The disease was reported from Honduras in 1972, from where it spread north and south from Mexico to Brazil and into the Caribbean islands, in 1991. The fungus arrived in Zambia in 1973 and spread to the banana-producing areas of Africa from that introduction. The first occurrence of black sigatoka in Florida was reported in 1999. As it spread, black sigatoka replaced the yellow form and has become the dominant disease of bananas worldwide.

The most likely route of infection is through the importation of infected plant material, and infection can spread rapidly in commercial areas where bananas are farmed in monoculture.

In commercial export plantations, the disease causes up to 50% loss of fruit and is controlled only by frequent applications of fungicides. Removal of affected leaves, good drainage, and sufficient spacing also help to fight the disease. Although fungicides improved over the years, the pathogen developed resistance. Therefore, higher frequency of applications is required, increasing the impact on the environment and health of the banana workers. In regions where disease pressure is low and fungicide resistance has not been observed, it is possible to better time the application of systemic fungicides by using a biological forecasting system.

Small farmers growing bananas for local markets cannot afford expensive measures to fight the disease. However, some cultivars of bananas are resistant to the disease. Research is done to improve productivity and fruit properties of these cultivars. A genetically modified banana variety made more resistant to the fungus was developed and was field tested in Uganda in the late 2000s.

# Leaf Rust (Barley)

Leaf rust is a fungal disease of barley caused by *Puccinia hordei*. It is also known as brown rust and it is the most important rust disease on barley.

## Symptoms

Pustules of leaf rust are small and circular, producing a mass of orange-brown powdery spores. They appear on the leaf sheaths and predominantly on the upper leaf surfaces. Heavily infected leaves die prematurely.

## Crop Losses

Leaf rust of barley is considered a relatively minor disease in the United States. However, sporadic outbreaks have occurred in the southeastern and Midwestern regions of the country.

## Pathotypes and Host Resistance

Most of the barley cultivars grown in the United States are susceptible to *Puccinia hordei*. Nineteen seedling resistance genes (i.e. *Rph1* to *Rph19*) have been identified, but only three (*Rph3, 7* and *9*) have been deployed in commercial cutlivars worldwide. In the United States, the *Rph7* gene effectively controlled the disease for over twenty years. However, in 1993, pathotypes with virulence to the *Rph7* resistance gene were identified in Virginia, California, and Pennsylvania. Recently, the first simply inherited gene conferring adult plant resistance to leaf rust in barley was designated *Rph20*. *Rph20* originated from the two-rowed barley landrace *H. laevigatum* (i.e., *Hordeum vulgare* subsp. *vulgare*); parent of the Dutch cultivar 'Vada' (released in the 1950s). To date there have been no reports of an *Rph20*-virulent pathotype.

# Corn Grey Leaf Spot

Grey leaf spot (GLS) is a foliar fungal disease that affects maize, also known as corn. There are two fungal pathogens that cause GLS, which are *Cercospora zeae-maydis* and *Cercospora zeina* . Symptoms seen on corn include leaf lesions, discoloration (chlorosis), and foliar blight. The fungus survives in debris of topsoil and infects healthy crop via asexual spores called conidia. Environmental conditions that best suit infection and growth include moist, humid, and warm climates. Poor airflow, low sunlight, overcrowding, improper soil nutrient and irrigation management, and poor soil drainage can all contribute to the propagation of the disease. Management techniques include crop resistance, crop rotation, residue management, use of fungicides, and weed control. The purpose of disease management is to prevent the amount of secondary disease cycles as well as to protect leaf area from damage prior to grain formation. Corn grey leaf spot is an important disease of corn production in the United States, economically significant throughout the Midwest and Mid-Atlantic regions. However, it is also prevalent in Africa, Central America, China, Europe, India, Mexico, the Philippines, northern South America, and Southeast Asia. The teleomorph (sexual phase) of *Cercospora Zeae-Maydis* is assumed to be *Mycosphaerella sp.*

## Host and Symptoms

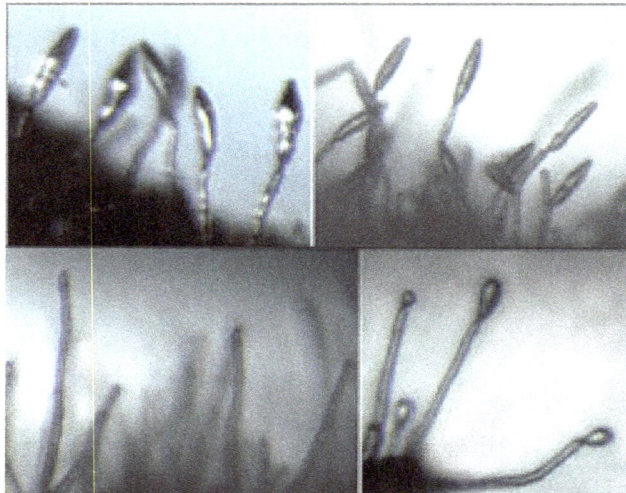

Conidiophores of corn grey leaf spot

Corn is the only species that can be affected by *Cercospora zeae-maydis*. There are two populations of *Cercospora zeae-maydis*, distinguished by molecular analysis, growth rate, geographic distribution, and cercosporin toxin production. *Cercospora Zeae-Maydis* differs from its cousin group *Cercospera zeina sp. nov* in that it has faster growth rate in artificial media, the ability to produce the toxin cercosporin, longer conidiophores, and broadly fusiform conidia. *Cercospera zeina sp. nov* affects corn in the Eastern Corn Belt and Mid-Atlantic States; *Cercospora Zeae-May-*

*dis* is found in most corn producing areas of western Kentucky, Illinois, Indiana, Iowa, Wisconsin, Missouri, Ohio, and west Tennessee (Midwest). Both populations share the same symptoms and virulence, the ability of the fungus to invade the host.

Major outbreaks of grey leaf spot occur whenever favorable weather conditions are present . The initial symptoms of grey leaf spot emerge as small, dark, moist spots that are encircled by a thin, yellow radiance (lesions forming). The tissue within the "spot" begins to die as spot size increases into longer, narrower leaf lesions. Although initially brownish and yellow, the characteristic grey color that follows is due to the production of grey fungal spores (conidia) on the lesion surface. These symptoms that are similar in shape, size and discoloration, are also prevalent on the corn husks and leaf sheaths. Leaf sheath lesions are not surrounded by a yellow radiance, rather a brown or dark purple radiance. This dark brown or purple discoloration on leaf sheaths is also characteristic to northern corn leaf blight (*Exserohilum turcicum*), southern corn leaf blight (*Bipolaris maydis*), or northern corn leaf spot (*Bipolaris zeicola*). Corn grey leaf spot mature lesions are easily diagnosed and distinguishable from these other diseases. Mature corn grey leaf spot lesions have brown rectangular and vein limited shape. Secondary and tertiary leaf veins limit the width of the lesion and sometimes individual lesions can combine to blight entire leaves.

## Pathogenesis

One reason for the pathogenic success of *Cercospora zeae-maydis* is the production of a plant toxin called cercosporin. All members of the *Cercospora* genus produce this light-activated toxin during infection. In the absence of light, cercosporin is inactive, but when light is present, the toxin is converted into its excited triplet state. Activated cercosporin reacts with oxygen molecules, generating active single oxygen radicals. Oxygen radicals react with plant cell lipids, proteins, and nucleic acids, damaging and killing affected cells, and nutrients released during the cell rupture and death feed the *Cercospora* fungus. A study of mutant *Cercospora* lacking the gene responsible for cercosporin production demonstrates that, though unnecessary for infection, cercosporin increases the virulence of *Cercospora* fungi.

## Disease Cycle

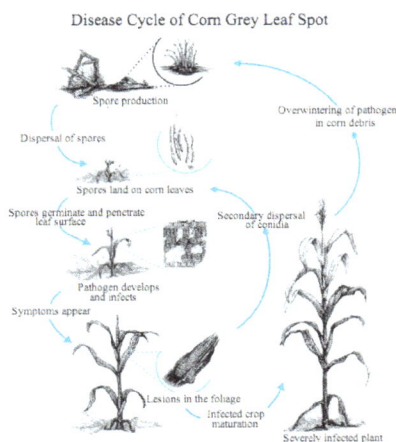

Life cycle of corn grey leaf spot

*Cercospora zeae-maydis* survives only as long as infected corn debris is present; however, it is a poor soil competitor. The debris on the soil surface is a cause for primary inoculation that infects the incoming corn crop for the next season. By late spring, conidia (asexual spores) are produced by *Cercospora zeae-maydis* in the debris through wind dispersal or rain. The conidia are disseminated and eventually infect new corn crop. In order for the pathogen to actually infect the host, high relative humidity and moisture (dew) on the leaves are necessary for inoculation. Primary inoculation occurs on lower regions of younger leaves, where conidia germinate across leaf surfaces and penetrate through stomata via a flattened hyphal organ, an appressorium. *Cercospora zeae-maydis* is atypical in that its conidia can grow and survive for days before penetration, unlike most spores that need to penetrate within hours to ensure survival. Once infection occurs, the conidia are produced in these lower leaf regions. Assuming favorable weather conditions, these conidia serve as secondary inoculum for upper leaf regions, as well as husks and sheaths (where it can also overwinter and produce conidia the following season). Additionally, wind and heavy rains tend to disperse the conidia during many secondary cycles to other parts of the field causing more secondary cycles of infection. If conditions are unfavorable for inoculation, the pathogen undergoes a state of dormancy during the winter season and reactivates when conditions favorable to inoculation return (moist, humid) the following season. The fungus overwinters as stromata (mixture of plant tissues and fungal mycelium) in leaf debris, which give rise to conidia causing primary inoculations the following spring and summer.

## Environment

Corn grey leaf spot flourishes under extended periods of high relative humidity (> 2 days) and free moisture on leaves due to fog, dew, or light rain. Additionally, heavy rains tend to assist in dispersal of the pathogen. Temperatures between 75° and 95 °F are also required. If temperature drops below 75 °F during wet periods or lack 12 hours of wetness, the extent of disease will be greatly diminished. In the Midwest and Mid-Atlantic, these conditions are favorable for spore development during the spring and summer months.

## Management

In order to best prevent and manage corn grey leaf spot, the overall approach is to reduce the rate of disease growth and expansion. This is done by limiting the amount of secondary disease cycles and protecting leaf area from damage until after corn grain formation. High risks for corn grey leaf spot are divided into eight factors, which require specific management strategies.

High risk factors for grey leaf spot in corn:

1. Susceptible hybrid

2. Continuous corn

3. Late planting date

4. Minimum tillage systems

5. Field history of severe disease

6.  Early disease activity (before tasseling)

7.  Irrigation

8.  Favorable weather forecast for disease

There are currently five different management strategies, some of which are more effective than others.

## Resistant Varieties

The most proficient and economical method to reduce yield losses from corn grey leaf spot is by introducing resistant plant varieties. In places where leaf spot occur, these crops can ultimately grow and still be resistant to the disease. Although the disease is not completely eliminated and resistant varieties show disease symptoms, at the end of the growing season, the disease is not as effective in reducing crop yield. SC 407 have been proven to be common corn variety that are resistant to grey leaf spot. If grey leaf spot infection is high, this variety may require fungicide application to achieve full potential. Susceptible varieties should not be planted in previously infected areas.

## Crop Rotation

The amount of initial inoculum will be reduced when a crop other than corn is planted for ≥2 years in that given area; meanwhile proper tillage methods are carried out. Clean plowing and 1-year crop rotation in the absence of corn allows for greater reductions of the disease as well. Note that conventional tilling can reduce disease but can lead to greater soil erosion.

## Residue Management

Burying the debris under the last year's crop will help in reducing the presence of *Cercospera zeae-maydis*, as the fungal-infected debris can only survive above the soil surface. Again this technique will aid in reducing the primary inoculum, but it will not completely eradicate the disease.

## Fungicides

Fungicides, if sprayed early in season before initial damage, can be effective in reducing disease.

Currently there are 5 known fungicides that treat Corn grey leaf spot:

1.  Headline EC (active ingredient: pyraclostrobin)

2.  Quilt (active ingredient: azoxystrobin + propiconazole)

3.  Proline 480 SC (active ingredient: prothioconazole)

4.  Tilt 250 E, Bum per 418 EC (active ingredient: propiconazole)

    Headline EC

    Headline is to be applied at 400-600 mL per/hectacre (ha). For optimal disease control, begin applications prior to disease development. This fungicide can only be applied a maximum of 2 applications/year. Ground and aerial application are both acceptable.

Quilt

Quilt is to be applied at 0.75-1.0 L per/ha. Application of Quilt is to be made upon first appearance of disease, followed by a second application 14 days after, if environmental conditions are favorable for disease development. Upon browning of corn sheaths, Quilt is not to be applied. This fungicide can only be applied a maximum 2 applications/yr. Ground and aerial application are both acceptable.

Proline 480 SC

Proline 480 SC is to be applied at 420 mL per/ha. This fungicide can only be applied a maximum 1 time/year. It should be note that only ground application is acceptable. A 24-hour re-entry time is required (minimum amount of time that must pass between the time a fungicide is applied to an area or crop and the time that people can go into that area without protective clothing and equipment).

Tilt 250 and Bumper 418 EC

Tilt 250 is to be applied at 500 mL per/ha. Bumper 418 EC is to be applied at 300 mL per/ha. Both fungicides are to be applied when rust pustules first appear. If disease is prevalent after primary application, a second application 14 days later may be necessary. Two weeks later, a third application can be made under severe amount of disease. Ground and aerial application are both acceptable.

When spraying fungicides Quilt and Headline EC at 6 oz/a at tassel stage using a tractor-mounted $CO_2$ powered sprayer using 20 gallons of water/acre, average yield was seen to increase. The use of fungicides can be both economically and environmentally costly and should only be applied on susceptible varieties and large-scale corn production. In order to prevent fungal resistance to fungicides, all fungicides are to be used alternatively, switching fungicides with different modes of action. Pyraclostrobin (Headline EC) and azoxystrobin are Quinone outside Inhibitor (QoI) fungicides, whereas propiconazole and prothioconazole are DeMethylation Inhibitors (DMI) fungicides.

## Weed Control

By removing weeds, above ground airflow to the crop is increased, relative humidity is decreased, and it limits infection at most susceptible times.

## Importance

Before 1970, corn grey leaf spot was not prevalent in the United States, however the disease spread during the mid part of the decade throughout low mountain regions of North Carolina, Kentucky, Tennessee, and Virginia. Today, the disease has expanded to Delaware, Illinois, Indiana, Iowa, Maryland, Missouri, Ohio, Pennsylvania and west Tennessee. Corn grey leaf spot can be an extremely devastating disease as potential yield losses range from 5 to 40 bushels/acre. At higher disease levels, even greater losses can result. When a corn plant's ability to store and produce carbohydrates (glucose) in the grain is diminished, yield losses take place. This occurs when *Cercospera zeae maydis* infects foliar tissue and reduces the plant's ability to photosynthesize and produce byproducts of the process (ex. glucose).

## References

- Singh, Prof. V.; Dr. P. C. Pandey; Dr. D. K. Jain (2008). A Text Book of Botany. India: Rastogi. p. 15.132. ISBN 978-81-7133-904-4.

- Meredith, D.S. (1 January 1970). Banana Leaf Spot Disease (Sigatoka) Caused by Mycosphaerella Musicola Leach. Commonwealth Mycological Institute, Kew, Surrey, England. ISBN 978-0-00-000089-7. Retrieved 13 August 2013.

- "Biological forecasting system for black leaf streak — Knowledge and news on bananas from ProMusa". Retrieved 29 March 2016.

- Hsiang, T.; Goodwin, P.H. (July 2001). "Ribosomal DNA Sequence Comparisons of Cholletotrichum Graminicola from Turfgrasses and other Hosts". European Journal of Plant Pathology. 107 (6): 593–599. doi:10.1023/A:1017974630963#page-1 (inactive 2016-01-07). Retrieved November 11, 2015.

- Daub, Margaret; Chung, Kuang-Ren (2007). "Cercosporin: A Photoactivated Toxin in Plant Disease". APSnet Features. doi:10.1094/APSnetFeature/2007-0207. Retrieved October 20, 2015.

- Jones, David Robert [editor] (2000). Diseases of Banana, Plantain, Abaca and Enset. Wallingford, Oxon, UK: CABI Publishing. pp. 79–92. OCLC 41347037. Retrieved 13 August 2013.

- Ploetz, R.C. (2001). "Black sigatoka of banana: The most important disease of a most important fruit". The Plant Health Instructor. doi:10.1094/PHI-I-2001-0126-02. Retrieved 13 August 2013.

# Epidemiology and Morphology of Plant Diseases

Plant disease epidemiology is the study of disease in plant pollutions. Plant diseases occur due to pathogens such as bacteria, viruses, fungi and parasitic plants. Early detection and accurate diagnosis is essential for effective management of plant disease. This chapter provides a comprehensive overview of plant disease.

## Plant Disease Epidemiology

Plant disease epidemiology is the study of disease in plant populations. Much like diseases of humans and other animals, plant diseases occur due to pathogens such as bacteria, viruses, fungi, oomycetes, nematodes, phytoplasmas, protozoa, and parasitic plants. Plant disease epidemiologists strive for an understanding of the cause and effects of disease and develop strategies to intervene in situations where crop losses may occur. Typically successful intervention will lead to a low enough level of disease to be acceptable, depending upon the value of the crop.

Plant disease epidemiology is often looked at from a multi-disciplinary approach, requiring biological, statistical, agronomic and ecological perspectives. Biology is necessary for understanding the pathogen and its life cycle. It is also necessary for understanding the physiology of the crop and how the pathogen is adversely affecting it. Agronomic practices often influence disease incidence for better or for worse. Ecological influences are numerous. Native species of plants may serve as reservoirs for pathogens that cause disease in crops. Statistical models are often applied in order to summarize and describe the complexity of plant disease epidemiology, so that disease processes can be more readily understood. For example, comparisons between patterns of disease progress for different diseases, cultivars, management strategies, or environmental settings can help in determining how plant diseases may best be managed. Policy can be influential in the occurrence of diseases, through actions such as restrictions on imports from sources where a disease occurs.

In 1963 J. E. van der Plank published "Plant Diseases: Epidemics and Control", a seminal work that created a theoretical framework for the study of the epidemiology of plant diseases. This book provides a theoretical framework based on experiments in many different host pathogen systems and moved the study of plant disease epidemiology forward rapidly, especially for fungal foliar pathogens. Using this framework we can now model and determine thresholds for epidemics that take place in a homogeneous environment such as a mono-cultural crop field.

### Elements of An Epidemic

Disease epidemics in plants can cause huge losses in yield of crops as well threatening to wipe out an entire species such as was the case with Dutch Elm Disease and could occur with Sudden Oak Death. An epidemic of potato late blight, caused by *Phytophthora infestans*, led to the Great Irish Famine and the loss of many lives.

Commonly the elements of an epidemic are referred to as the "disease triangle": a susceptible host, pathogen, and conducive environment. For disease to occur all three of these must be present. Below is an illustration of this point. Where all three items meet there is disease. The fourth element missing from this illustration for an epidemic to occur, is time. As long as all three of these elements are present disease can initiate, an epidemic will only ensue if all three continue to be present. Any one of the three might be removed from the equation though. The host might out-grow susceptibility as with high-temperature adult-plant resistance, the environment changes and is not conducive for the pathogen to cause disease, or the pathogen is controlled through a fungicide application for instance.

Sometimes a fourth factor of time is added as the time at which a particular infection occurs, and the length of time conditions remain viable for that infection, can also play an important role in epidemics. The age of the plant species can also play a role, as certain species change in their levels of disease resistance as they mature; a process known as ontogenic resistance.

If all of the criteria are not met, such as a susceptible host and pathogen are present but the environment is not conducive to the pathogen infecting and causing disease, disease cannot occur. For example, corn is planted into a field with corn residue that has the fungus *Cercospora zea-maydis*, the causal agent of Grey leaf spot of corn, but if the weather is too dry and there is no leaf wetness the spores of the fungus in the residue cannot germinate and initiate infection.

Likewise, it stands to reason if the host is susceptible and the environment favours the development of disease but the pathogen is not present there is no disease. Taking the example above, the corn is planted into a ploughed field where there is no corn residue with the fungus *Cercospora zea-maydis*, the causal agent of Grey leaf spot of corn, present but the weather means long periods of leaf wetness, there is no infection initiated.

When a pathogen requires a vector to be spread then for an epidemic to occur the vector must be plentiful and active.

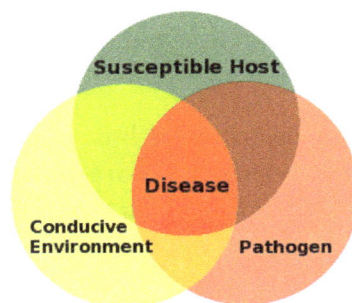

Plant disease triangle illustration

## Types of Epidemics

Monocyclic epidemics are caused by pathogens with a low birth rate and death rate, meaning they only have one infection cycle per season. They are typical of soil-borne diseases such as Fusarium wilt of flax. Polycyclic epidemics are caused by pathogens capable of several infection cycles a season. They are most often caused by airborne diseases such as powdery mildew. Bimodal polycyclic epidemics can also occur. For example, in brown rot of stone fruits the blossoms and the fruits are infected at different times.

For some diseases it is important to consider the disease occurrence over several growing seasons, especially if growing the crops in monoculture year after year or growing perennial plants. Such conditions can mean that the inoculum produced in one season can be carried over to the next leading to a build of an inoculum over the years. In the tropics there are no clear-cut breaks between growing seasons as there are in temperate regions and this can lead to accumulation of inoculum.

Epidemics that occur under these conditions are referred to as *polyetic* epidemics and can be caused by both monocyclic and polycyclic pathogens. Apple powdery mildew is an example of a polyetic epidemic caused by a polycyclic pathogen and Dutch Elm disease a polyetic epidemic caused by a monocyclic pathogen.

### Employment

Plant disease epidemiologists are typically employed as researchers by universities, or governmental institutions such as the USDA. However, private companies in agricultural fields also employ epidemiologists.

## Morphological Symptoms of Plant Diseases

Thousands of plant diseases have been recorded throughout the world, many of these causing heavy crop loses. Early detection and accurate diagnosis is essential for the effective management of plant disease. Thus the first step in studying any disease is its timely detection of the diseased plant. Quick initial detection is largely based on the signs and symptoms of disease.

Signs are the visible physical presence of either the pathogen itself or the structures formed by the pathogen. Common examples of easily detected signs are those such as the fungal mycelia and spore masses of downy mildews observed on infected leaves and the bacterial ooze of *Xanthomonas* leaf streak disease on rice.

Symptoms are the visible changes that occur in the host plant in response to infection by pathogens. For any disease in a given plant, there is the characteristic expression of symptoms, usually occurring in a sequential series during the course of the disease. This series of symptoms depicting the disease picture is referred to as the disease syndrome.

Morphological symptoms may be exhibited by the entire plant or by any organ of the plant. These have been categorized into different groups for easy of study. Primarily, morphological symptoms of plant diseases can be categorized into 6 different types.

- Necroses
- Growth abnormalities

- Metaplastic symptoms
- Proleptic symptoms
- Color changes
- Wilts

## Necroses

Necroses are caused due to necrosis or death of plant cells. The affected plant tissue usually turns brown to black in color. Necrotic symptoms could appear in any part of the plant such as in storage organs, in green tissues, or in woody tissues.

## Necrosis in Storage Organs

Death of cells in storage organs terminates in decomposition or decay referred to as a rot. Two types of rots are identified as Dry rot and Wet rot on storage tissues.

Soft rots are those where the pathogen breaks down the host cell walls, resulting in the exudation of juices from the infected tissue. The organ becomes mushy or pulpy and a foul smell often develops due to colonization by secondary invaders. Many fungi and bacteria cause soft rots on several fruits and vegetables. Species of the fungus, *Rhizopus* and bacterium *Erwinia* are two such commonly found pathogens causing soft rots.

In a dry rot, the storage organ becomes hard and dry, and in some diseases, there is rapid loss of water and the infected organs become shriveled, wrinkled, and leathery. Dry rots showing such symptoms are referred to as mummifications.

## Necrosis in Green Tissues

Necroses on green tissue are termed differently based on the nature of symptoms and the type of green tissue. The term, damping off refers to the sudden wilting and topping over of seedlings as a result of extensive necrosis of tender tissue of the roots and stem near the soil line, due to the attack of soil-borne pathogens such as fungus, *Pythium*. This fungus is known to cause damping off in an assortment of seedlings such as that of brinjal, chilli, mung beans, tobacco, tomato, and *Cucurbita*.

A spot refers to a well-defined area of gray or brown necrotic tissue. Spots are very common on leaves and fruits and are probably the most familiar necrotic symptom. Sometimes the necrotic tissue within a leaf spot may crack and fall off from the surrounding green tissue leaving an empty space. Such a symptom is known as a shot hole.

Minute or very small spots are sometimes referred to as flecks or specks.

When dark mycelia of a fungal pathogen appear on the surface of necrotic spot, blotting the leaves, shoots, an stems as large and irregular spots, the symptom is referred to as a blotch.

Both streaks and stripes occur in grasses and are elongated areas having dead cells. Streaks occur along the stem and veins, while stripes are in the laminar tissues between veins.

Net necrosis is a symptom resulting from an irregular pattern of anastomoses between streaks or stripes.

Blights are characterized by the rapid death of entire leaves including the veins or parts of the leaves. Blights also could occur on flowers and stems.

Scorches resemble blights, but there necrosis occurs in irregular patterns between veins and along leaf margins.

Firing is sudden drying, collapse and death of entire leaves. Firing occurs in response to the activity of root rot and vascular wilt pathogens.

Scald is the term used to describe the blanching of epidermal and adjacent tissues of fruits and occasionally of leaves.

The sudden death of unopened buds or inflorescence is referred to as blast.

Extensive necrosis of fruits that resemble in premature dropping is called shelling.

## Necrosis in Woody Tissues

Necrosis of woody tissue often brings about various types of die-back symptoms. Dieback is the extensive necrosis of a shoot from its tip downwards.

Restricted necrosis of the bark and cortical tissue of stems and roots is termed as a canker. In cankers, necrotic tissue in the sunken lesions is sharply limited, usually by a callus from adjacent healthy tissue.

When woody tissues are diseased, they may exude different kinds of substances. When the exudate is gummy, the symptom is called gummosis, while it is resinosis when it is resinous. If the exudate is neither gummy nor resinous, it is described as bleeding.

## Abnormalities in Growth

Many disease symptoms are associated with growth changes in diseased plants. These could be caused by either reduced growth due to hypoplasia and atrophy or excessive growth due to hyperplasia and hypertrophy.

## Hypoplasia & Atrophy

Hypoplasia is the failure of plants or plant organs to develop fully due to a decreased production of the number of cells. Hypoplasia results in plants or plant parts of sub-normal size.

Atrophy is the reduction in the size of plant cells produced. This also results in stunted plants or plant parts.

Dwarfing is the failure of a plant or a plant part to attain its full size.

Rosetting is a condition where the internode of a plant do not elongate, and hence, the leaves appear close together in a cluster.

## Hyperplasia & Hypertrophy

Hyperplasia is the enlargement of a plant tissue due to excessive increase in the number of plant cells produced. Hyperplasia results in overdevelopment in size of plants or plant organs.

Hypertrophy is excessive growth due to the enlargement of individual cells. This condition also results in the overdevelopment in size of plants or plant organs.

Hyperplasia and hypertrophy could result in the enlargement of leaves and fruits, and the enlargement of stems and roots.

## Enlargement of Leaves and Fruits

Several symptoms expressing enlargement of leaves and fruits are commonly observed among diseased plants.

Curling which is the bending of the shoot or the rolling of the leaf, is a result of over-growth on one side of an organ. Often viral diseases cause such leaf distortions due to irregular growth of the lamina. Extreme reduction of the leaf lamina brings about the symptom known as the Shoe-string effect.

The puckering or crinkling of leaves due to different growth rates in adjacent tissue is known as savoying.

Overgrowth of epidermal and underlying tissues of leaves, stems, fruits and tubers may result scab formation. Scab consists of raised, rough, and discrete lesions. These are often sunken and cracked, giving a typical scabby appearance.

Localized swellings or enlargement of epidermal cells due to excessive accumulation of water is termed intermuscence and the diagnostic symptom is the appearance of a blister.

## Enlargement of Stems and Roots

Symptoms causing enlargement of stems and roots are termed differently based on their nature. Excessive accumulation of food material in stems, above a constricted area produces a swelling termed sarcody.

Localized swellings that involve entire organs are termed tumefaction. Commonly exhibited tumefactions are galls, clubs, and knots.

Excessive development of adventitious organs results in fasciculation, that is the clustering of organs around a focal point. Such examples include witch's broom and hairy root. Witch's broom is a broom-like mass proliferation due to the dense clustering of branches of woody plants while hairy root results due to excessive development of roots.

Fasciation is the broadening or flattening of cylindrical organs such as stems. The continued development of any organ after it has reached a stage beyond which it normally does not grow is known as proliferation.

The outgrowth of tissue in response to wounding is known as a callus. Callus formation is found to form around most cankers.

## Metaplastic symptoms

Metaplastic symptoms are those which form when tissues change from one form to another. Such symptoms include phyllody, the development of floral organs into leaf-like structures, juvenillody, the development of juvenile seedlings on mature plants and russeting, a superficial browning of surfaces of fruits and tubers due to suberization.

## Proleptic Symptoms

Proleptic symptoms result from the development of tissues earlier than usual. Examples include prolepsis, the premature development of a shoot from a bud, proleptic abscission, the premature formation of abscission layers and restoration, the unexpected development of organs that are normally rudimentary.

## Color Changes

Changes in the color of plant tissue are a common symptom of plant disease. Often these color changes are brought about by the yellowing of normal green tissue due to the destruction of chlorophyll or a failure to form chlorophyll. Such repression of leaf color may be complete or partial.

When color repression is complete, it is known as albication. However, the more common, partial repression is referred to as chlorosis.

Patches of green tissue alternating with chlorotic areas are described as a mosaic. Mosaic is a symptom caused by many viruses. Based on the intensity and the pattern of discoloration, mosaics are termed differently. Irregular patches of distinct light and dark areas are known as mottling. Streaking and ring spots are still other distinct types of discolorations. Ring spots are circular masses of chlorosis with a green center. Vein clearing and vein banding are yet other common color changes on leaves.

Chlorophyll may also develop in tissues normally devoid of it. Thus usually white or colored tissue becomes green in color. This is called as virescence.

Anthocyanescence is due to the overdevelopment of anthocyanin and result in the development of a purplish coloration. Color changes can also take place in flowers. Such an example is the color break virus-affected tulips.

## Wilts

Wilting is due to loss of turgor in plant tissue resulting in the dropping of plant parts. They are common symptom in diseases where the pathogen or the toxic metabolites it produces affects the vascular tissue of the host plant. Interference in water transport brought about by the infection of these vascular pathogens leads to wilting. Unlike wilting due to low soil moisture, wilting due to the activity of these pathogens cannot be overcome by watering the plants. Infected plants eventually die.

# Protection of Plants from Diseases

Protecting plants from diseases is very important and this chapter explains the various protections of plants from diseases. Plant disease resistance, diseases resistance in fruit and vegetables, pest control and fungicide are some of the protections, which are discussed in the following content.

## Plant Disease Resistance

Plant disease resistance protects plants from pathogens in two ways: mechanisms and by infection-induced responses of the immune system. Relative to a susceptible plant, disease resistance is the reduction of pathogen growth on or in the plant, while the term disease tolerance describes plants that exhibit little disease damage despite substantial pathogen levels. Disease outcome is determined by the three-way interaction of the pathogen, the plant and the environmental conditions (an interaction known as the disease triangle).

Cankers caused Chestnut blight a disease that affects the chestnut tree

Defense-activating compounds can move cell-to-cell and systemically through the plant vascular system. However, plants do not have circulating immune cells, so most cell types exhibit a broad suite of antimicrobial defenses. Although obvious *qualitative* differences in disease resistance can be observed when multiple specimens are compared (allowing classification as "resistant" or "susceptible" after infection by the same pathogen strain at similar inoculum levels in similar environments), a gradation of *quantitative* differences in disease resistance is more typically observed between plant strains or genotypes. Plants consistently resist certain pathogens but succumb to others; resistance is usually pathogen species- or pathogen strain-specific.

## Background

Plant disease resistance is crucial to the reliable production of food, and it provides significant reductions in agricultural use of land, water, fuel and other inputs. Plants in both natural and cultivated populations carry inherent disease resistance, but this has not always protected them.

The late blight Irish potato famine of the 1840s was caused by the oomycete Phytophthora infestans. The world's first mass-cultivated banana cultivar Gros Michel was lost in the 1920s to Panama disease caused by the fungus Fusarium oxysporum. The current wheat stem, leaf, and yellow stripe rust epidemics spreading from East Africa into the Indian subcontinent are caused by rust fungi Puccinia graminis and P. striiformis. Other epidemics include Chestnut blight, as well as recurrent severe plant diseases such as Rice blast, Soybean cyst nematode, Citrus canker.

Plant pathogens can spread rapidly over great distances, vectored by water, wind, insects, and humans. Across large regions and many crop species, it is estimated that diseases typically reduce plant yields by 10% every year in more developed nations or agricultural systems, but yield loss to diseases often exceeds 20% in less developed settings, an estimated 15% of global crop production.

However, disease control is reasonably successful for most crops. Disease control is achieved by use of plants that have been bred for good resistance to many diseases, and by plant cultivation approaches such as crop rotation, pathogen-free seed, appropriate planting date and plant density, control of field moisture and pesticide use.

## Viral Disease Common Mechanisms

## Pre-formed Structures and Compounds

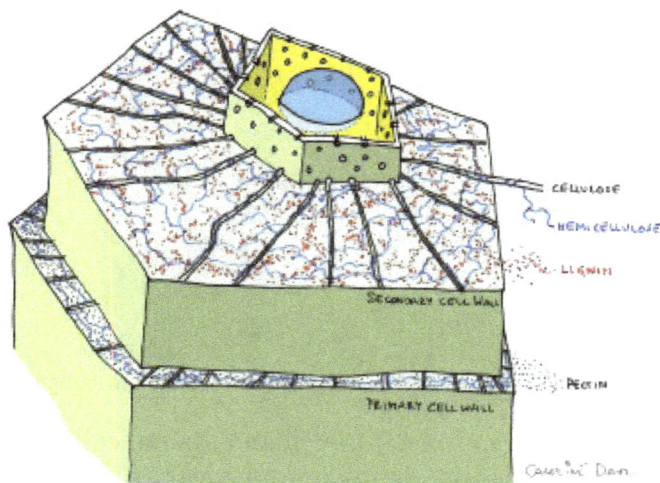

secondary plant wall

- Plant cuticle/surface

- Plant cell walls

- Antimicrobial chemicals (for example: glucosides, saponins)

- Antimicrobial proteins

- Enzyme inhibitors

- Detoxifying enzymes that break down pathogen-derived toxins

- Receptors that perceive pathogen presence and activate inducible plant defences

## Inducible Post-infection Plant Defenses

- Cell wall reinforcement (callose, lignin, suberin, cell wall proteins)

- Antimicrobial chemicals, including reactive oxygen species such as hydrogen peroxide or peroxynitrite, or more complex phytoalexins such as genistein or camalexin

- Antimicrobial proteins such as defensins, thionins, or PR-1

- Antimicrobial enzymes such as chitinases, beta-glucanases, or peroxidases

- Hypersensitive response - a rapid host cell death response associated with defence mediated by "Resistance genes."(Bryant, Tracy 2008).

## Variable Resistance

Even in susceptible plants to obligate parasites the tissue resistance changes due to ontogeny and to influence of external conditions. This resistance can be measured by the value of redox potential of electron carriers, which is produced in the plant by enzymatic reactions associated with respiration. Electron carriers are water-soluble and are not oxidized by air oxygen. There is not free oxygen in the cells, all the oxidizations and reductions take place enzymatically. These reactions are highly specific for the plant species. The host and parasite have different electron carriers.

## Immune System

The plant immune system consists of two interconnected tiers of receptors, one outside and one inside the cell. Both systems sense the intruder, respond to the intrusion and optionally signal to the rest of the plant and sometimes to neighboring plants that the intruder is present. The two systems detect different types of pathogen molecules and classes of plant receptor proteins

The first tier is primarily governed by pattern recognition receptors that are activated by recognition of evolutionarily conserved pathogen or microbial–associated molecular patterns (PAMPs or MAMPs, here P/MAMP). Activation of PRRs leads to intracellular signaling, transcriptional reprogramming, and biosynthesis of a complex output response that limits colonization. The system is known as PAMP-Triggered Immunity (PTI)"(JonesDangl2010).

The second tier (again, primarily), effector-triggered immunity (ETI), consists of another set of receptors, (nucleotide-binding)They operate within the cell, encoded by R genes. The presence of specific pathogen "effectors" activates specific NLR proteins that limit pathogen proliferation.

Receptor responses include ion channel gating, oxidative burst, cellular redox changes, or protein kinase cascades that directly activate cellular changes (such as cell wall reinforcement or antimicrobial production), or activate changes in gene expression that then elevate other defensive responses

Plant immune systems show some mechanistic similarities with the immune systems of insects and mammals, but also exhibit many plant-specific characteristics. Plants can sense the presence of pathogens and the effects of infection via activated by touch]. Rice Universi

## PAMP-triggered Immunity

PAMP-Triggered Immunity conserved molecules that inhabit multiple pathogen genera are classified as MAMPs by some researchers. The defenses induced by MAMP perception are sufficient to repel most pathogens. However, pathogen effector proteins are adapted to suppress basal defenses such as PTI

## Effector Triggered Immunity

Effector Triggered Immunity (ETI) is activated by the presence of pathogen effectors. The ETI immune response is reliant on R genes, and is activated by specific pathogen strains. As with PTI, many specific examples of apparent ETI violate commoMost plant immune systems carry a repertoire of 100-600 different R genes that mediate resistance to various virus .Plant ETI often cause an apoptotic hypersensitive response.(Odds Rathjen2010).

## R Genes and R Proteins

Plants have evolved R genes (resistance genes) whose products allow recognition of specific pathogen effectors, either through direct binding or by recognition of the effector's alteration of a host protein. These virulence factors drove co-evolution of plant resistant genes to combat the pathogens' Avr (avirulent) genes. Many R genes encode NB-LRR proteins (nucleotide-binding/leucine-rich repeat domains, also known as NLR proteins or STAND proteins, among other names).

R gene products control a broad set of disease resistance responses whose induction is often sufficient to stop further pathogen growth/spread. Each plant genome contains a few hundred apparent R genes. Studied R genes usually confer specificity for particular pathogen strains. As first noted by Harold Flor in his mid-20th century formulation of the gene-for-gene relationship, the plant R gene and the pathogen Avr gene must have matched specificity for that R gene to confer resistance, suggesting a receptor/ligand interaction for Avr and R genes. Alternatively, an effector can modify its host cellular target (or a molecular decoy of that target) activating an NLR associated with the

## Effector Biology

So-called "core" effectors are defined operationally by their wide distribution across the population of a particular pathogen and their substantial contribution to pathogen virulence. Genomics can be used to identify core effectors, which can then functionally define new R alleles, which can serve as breeding targets.

## Rna Silencing And Systemic Acquired Resistance Elicited By Prior Infections

Against viruses, plants often induce pathogen-specific gene silencing mechanisms mediated by RNA interference. T

Plant immune systems also can respond to an initial infection in one part of the plant by physiologically elevating the capacity for a successful defense response in other parts. Such responses include systemic acquired resistance, largely mediated by salicylic acid-dependent pathways, and induced systemic resistance, largely mediated.

Species-level resistance

In a small number of cases, plant genes are effective against an entire pathogen species, even though that species that is pathogenic on other genotypes of that host species. Examples include barley MLO against powdery mildew, wheat Lr34 against leaf rust and wheat Yr36 against stripe rust. An array of mechanisms for this type of resistance may exist depending on the particular gene and plant-pathogen combination. Other reasons for effective plant immunity can include a lack of coadaptation (the pathogen and/or plant lack multiple mechanisms needed for colonization and growth within that host species), or a particularly effective suite of pre-formed defenses.

## Signaling Mechanisms

## Perception of Pathogen Presence

Plant defense signaling is activated by pathogen-detecting receptors. The activated receptors frequently elicit reactive oxygen and nitric oxide production, calcium, potassium and proton ion fluxes, altered levels of salicylic acid and other hormones and activation of MAP kinases and other specific protein kinases. These events in turn typically lead to the modification of proteins that control gene transcription, and the activation of defense-associated gene expression.

In addition to PTI and ETI, plant defenses can be activated by the sensing of damage-associated compounds (DAMP), such as portions of the plant cell wall released during pathogenic infection. Many receptors for MAMPs, effectors and DAMPs have been discovered. Effectors are often detected by NLRs, while MAMPs and DAMPs are often detected by transmembrane receptor-kinases that carry LRR or LysM extracellular domains.

## Transcription Factors and the Hormone Response

Numerous genes and/or proteins as well as other molecules have been identified that mediate plant defense signal transduction. Cytoskeleton and vesicle trafficking dynamics help to orient plant defense responses toward the point of pathogen attack.

## Mechanisms of Transcription Factors and Hormones

Plant immune system activity is regulated in part by signaling hormones such as:

- Salicylic acid

- Jasmonic acid

- Ethylene

There can be substantial cross-talk among these pathways.

## Regulation by Degradation

As with many signal transduction pathways, plant gene expression during immune responses can be regulated by degradation. This often occurs when hormone binding to hormone receptors stimulates ubiquitin-associated degradation of repressor proteins that block expression of certain genes. The net result is hormone-activated gene expression. Examples:

- Auxin: binds to receptors that then recruit and degrade repressors of transcriptional activators that stimulate auxin-specific gene expression.

- Jasmonic acid: similar to auxin, except with jasmonate receptors impacting jasmonate-response signaling mediators such as JAZ proteins.

- Gibberellic acid: Gibberellin causes receptor conformational changes and binding and degradation of Della proteins.

- Ethylene: Inhibitory phosphorylation of the EIN2 ethylene response activator is blocked by ethylene binding. When this phosphorylation is reduced, EIN2 protein is cleaved and a portion of the protein moves to the nucleus to activate ethylene-response gene expression.

### Ubiquitin and E3 Signaling

Ubiquitination plays a central role in cell signaling that regulates processes including protein degradation and immunological response. Although one of the main functions of ubiquitin is to target proteins for destruction, it is also useful in signaling pathways, hormone release, apoptosis and translocation of materials throughout the cell. Ubiquitination is a component of several immune responses. Without ubiquitin's proper functioning, the invasion of pathogens and other harmful molecules would increase dramatically due to weakened immune defenses.

### E3 Signaling

The E3 Ubiquitin ligase enzyme is a main component that provides specificity in protein degradation pathways, including immune signaling pathways. The E3 enzyme components can be grouped by which domains they contain and include several types. These include the Ring and U-box single subunit, HECT, and CRLs. Plant signaling pathways including immune responses are controlled by several feedback pathways, which often include negative feedback; and they can be regulated by De-ubiquitination enzymes, degradation of transcription factors and the degradation of negative regulators of transcription.

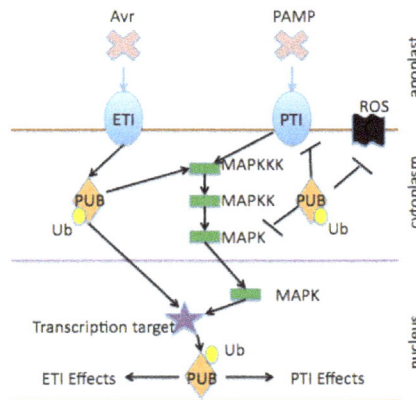

This image depicts the pathways taken during responses in plant immunity. It highlights the role and effect ubiquitin has in regulating the pathway.

## Plant Breeding for Disease Resistance

Plant breeders emphasize selection and development of disease-resistant plant lines. Plant diseases can also be partially controlled by use of pesticides and by cultivation practices such as crop rotation, tillage, planting density, disease-free seeds and cleaning of equipment, but plant varieties with inherent (genetically determined) disease resistance are generally preferred. Breeding for disease resistance began when plants were first domesticated. Breeding efforts continue because pathogen populations are under selection pressure for increased virulence, new pathogens appear, evolving cultivation practices and changing climate can reduce resistance and/or strengthen pathogens, and plant breeding for other traits can disrupt prior resistance. A plant line with acceptable resistance against one pathogen may lack resistance against others.

Breeding for resistance typically includes:

- Identification of plants that may be less desirable in other ways, but which carry a useful disease resistance trait, including wild strains that often express enhanced resistance.

- Crossing of a desirable but disease-susceptible variety to another variety that is a source of resistance.

- Growth of breeding candidates in a disease-conducive setting, possibly including pathogen inoculation. Attention must be paid to the specific pathogen isolates, to address variability within a single pathogen species.

- Selection of disease-resistant individuals that retain other desirable traits such as yield, quality and including other disease resistance traits.

Resistance is termed *durable* if it continues to be effective over multiple years of widespread use as pathogen populations evolve. "Vertical resistance" is specific to certain races or strains of a pathogen species, is often controlled by single R genes and can be less durable. Hoizontal or broad-spectrum resistance against an entire pathogen species is often only incompletely effective, but more durable, and is often controlled by many genes that segregate in breeding populations.

Crops such as potato, apple, banana and sugarcane are often propagated by vegetative reproduction to preserve highly desirable plant varieties, because for these species, outcrossing seriously disrupts the preferred traits. Asexual propagation. Vegetatively propagated crops may be among the best targets for resistance improvement by the biotechnology method of plant transformation to manage genes that affect disease resistance.

Scientific breeding for disease resistance originated with Sir Rowland Biffen, who identified a single recessive gene for resistance to wheat yellow rust. Nearly every crop was then bred to include disease resistance (R) genes, many by introgression from compatible wild relatives.

## GM or Transgenic Engineered Disease Resistance

The term GM ("genetically modified") is often used as a synonym of transgenic to refer to plants modified using recombinant DNA technologies. Plants with transgenic/GM disease resistance against insect pests have been extremely successful as commercial products, especially in maize and cotton, and are planted annually on over 20 million hectares in over 20 countries worldwide. Transgenic plant disease resistance against microbial pathogens was first demonstrated in 1986. Expression of viral coat protein gene sequences conferred virus resistance via small RNAs. This proved to be a widely applicable mechanism for inhibiting viral replication. Combining coat protein genes from three different viruses, scientists developed squash hybrids with field-validated, multiviral resistance. Similar levels of resistance to this variety of viruses had not been achieved by conventional breeding.

A similar strategy was deployed to combat papaya ringspot virus, which by 1994 threatened to destroy Hawaii's papaya industry. Field trials demonstrated excellent efficacy and high fruit quality. By 1998 the first transgenic virus-resistant papaya was approved for sale. Disease resistance has been durable for over 15 years. Transgenic papaya accounts for ~85% of Hawaiian production. The fruit is approved for sale in the U.S., Canada and Japan.

Potato lines expressing viral replicase sequences that confer resistance to potato leafroll virus were sold under the trade names NewLeaf Y and NewLeaf Plus, and were widely accepted in commercial production in 1999-2001, until McDonald's Corp. decided not to purchase GM potatoes and Monsanto decided to close their NatureMark potato business. NewLeaf Y and NewLeaf Plus potatoes carried two GM traits, as they also expressed Bt-mediated resistance to Colorado potato beetle.

No other crop with engineered disease resistance against microbial pathogens had reached the market by 2013, although more than a dozen were in some state of development and testing.

| Examples of transgenic disease resistance projects | | | | |
|---|---|---|---|---|
| Publication year | Crop | Disease resistance | Mechanism | Development status |
| 2012 | Tomato | Bacterial spot | R gene from pepper | 8 years of field trials |
| 2012 | Rice | Bacterial blight and bacterial streak | Engineered E gene | Laboratory |
| 2012 | Wheat | Powdery mildew | Overexpressed R gene from wheat | 2 years of field trials at time of publication |
| 2011 | Apple | Apple scab fungus | Thionin gene from barley | 4 years of field trials at time of publication |

| 2011 | Potato | Potato virus Y | Pathogen-derived resistance | 1 year of field trial at time of publication |
| --- | --- | --- | --- | --- |
| 2010 | Apple | Fire blight | Antibacterial protein from moth | 12 years of field trials at time of publication |
| 2010 | Tomato | Multibacterial resistance | PRR from *Arabidopsis* | Laboratory scale |
| 2010 | Banana | Xanthomonas wilt | Novel gene from pepper | Now in field trial |
| 2009 | Potato | Late blight | R genes from wild relatives | 3 years of field trials |
| 2009 | Potato | Late blight | R gene from wild relative | 2 years of field trials at time of publication |
| 2008 | Potato | Late blight | R gene from wild relative | 2 years of field trials at time of publication |
| 2008 | Plum | Plum pox virus | Pathogen-derived resistance | Regulatory approvals, no commercial sales |
| 2005 | Rice | Bacterial streak | R gene from maize | Laboratory |
| 2002 | Barley | Stem rust | Resting lymphocyte kinase (RLK) gene from resistant barley cultivar | Laboratory |
| 1997 | Papaya | Ring spot virus | Pathogen-derived resistance | Approved and commercially sold since 1998, sold into Japan since 2012 |
| 1995 | Squash | Three mosaic viruses | Pathogen-derived resistance | Approved and commercially sold since 1994 |
| 1993 | Potato | Potato virus X | Mammalian interferon-induced enzyme | 3 years of field trials at time of publication |

## PRR Transfer

Research aimed at engineered resistance follows multiple strategies. One is to transfer useful PRRs into species that lack them. Identification of functional PRRs and their transfer to a recipient species that lacks an orthologous receptor could provide a general pathway to additional broadened PRR repertoires. For example, the *Arabidopsis* PRR *EF-Tu* receptor (EFR) recognizes the bacterial translation elongation factor *EF-Tu*. Research performed at Sainsbury Laboratory demonstrated that deployment of EFR into either *Nicotiana benthamiana* or *Solanum lycopersicum* (tomato), which cannot recognize *EF-Tu*, conferred resistance to a wide range of bacterial pathogens. EFR expression in tomato was especially effective against the widespread and devastating soil bacterium Ralstonia solanacearum. Conversely, the tomato PRR *Verticillium 1* (*Ve1*) gene can be transferred from tomato to *Arabidopsis*, where it confers resistance to race 1 Verticillium isolates.

## Stacking

The second strategy attempts to deploy multiple NLR genes simultaneously, a breeding strategy known as stacking. Cultivars generated by either DNA-assisted molecular breeding or gene transfer will likely display more durable resistance, because pathogens would have to mutate multiple effector genes. DNA sequencing allows researchers to functionally "mine" NLR genes from multiple species/strains.

The *avrBs2* effector gene from *Xanthomona perforans* is the causal agent of bacterial spot disease of pepper and tomato. The first "effector-rationalized" search for a potentially durable R gene followed the finding that *avrBs2* is found in most disease-causing *Xanthomonas* species and is required for pathogen fitness. The *Bs2* NLR gene from the wild pepper, *Capsicum chacoense*, was moved into tomato, where it inhibited pathogen growth. Field trials demonstrated robust resistance without bactericidal chemicals. However, rare strains of *Xanthomonas* overcame *Bs2*-mediated resistance in pepper by acquisition of *avrBs2* mutations that avoid recognition but retain virulence. Stacking R genes that each recognize a different core effector could delay or prevent adaptation.

More than 50 loci in wheat strains confer disease resistance against wheat stem, leaf and yellow stripe rust pathogens. The Stem rust 35 (*Sr35*) NLR gene, cloned from a diploid relative of cultivated wheat, *Triticum monococcum*, provides resistance to wheat rust isolate *Ug99*. Similarly, *Sr33*, from the wheat relative *Aegilops tauschii*, encodes a wheat ortholog to barley *Mla* powdery mildew–resistance genes. Both genes are unusual in wheat and its relatives. Combined with the *Sr2* gene that acts additively with at least Sr33, they could provide durable disease resistance to *Ug99* and its derivatives.

## Executor Genes

Another class of plant disease resistance genes opens a "trap door" that quickly kills invaded cells, stopping pathogen proliferation. Xanthomonas and Ralstonia transcription activator–like (TAL) effectors are DNA-binding proteins that activate host gene expression to enhance pathogen virulence. Both the rice and pepper lineages independently evolved TAL-effector binding sites that instead act as an executioner that induces hypersensitive host cell death when up-regulated. *Xa27* from rice and Bs3 and Bs4c from pepper, are such "executor" (or "executioner") genes that encode non-homologous plant proteins of unknown function. Executor genes are expressed only in the presence of a specific TAL effector.

Engineered executor genes were demonstrated by successfully redesigning the pepper *Bs3* promoter to contain two additional binding sites for TAL effectors from disparate pathogen strains. Subsequently, an engineered executor gene was deployed in rice by adding five TAL effector binding sites to the *Xa27* promoter. The synthetic *Xa27* construct conferred resistance against Xanthomonas bacterial blight and bacterial leaf streak species.

## Host Susceptibility Alleles

Most plant pathogens reprogram host gene expression patterns to directly benefit the pathogen. Reprogrammed genes required for pathogen survival and proliferation can be thought of as "disease-susceptibility genes." Recessive resistance genes are disease-susceptibility candidates. For example, a mutation disabled an *Arabidopsis* gene encoding pectate lyase (involved in cell wall degradation), conferring resistance to the powdery mildew pathogen *Golovinomyces cichoracearum*. Similarly, the Barley *MLO* gene and spontaneously mutated pea and tomato *MLO* orthologs also confer powdery mildew resistance.

*Lr34* is a gene that provides partial resistance to leaf and yellow rusts and powdery mildew in wheat. *Lr34* encodes an adenosine triphosphate (ATP)–binding cassette (ABC) transporter. The dominant allele that provides disease resistance was recently found in cultivated wheat (not in wild strains) and, like *MLO* provides broad-spectrum resistance in barley.

Natural alleles of host translation elongation initiation factors *eif4e* and *eif4g* are also recessive viral-resistance genes. Some have been deployed to control potyviruses in barley, rice, tomato, pepper, pea, lettuce and melon. The discovery prompted a successful mutant screen for chemically induced *eif4e* alleles in tomato.

Natural promoter variation can lead to the evolution of recessive disease-resistance alleles. For example, the recessive resistance gene *xa13* in rice is an allele of *Os-8N3*. *Os-8N3* is transcriptionally activated by*Xanthomonas oryzae pv. oryzae* strains that express the TAL effector *PthXo1*. The *xa13* gene has a mutated effector-binding element in its promoter that eliminates *PthXo1* binding and renders these lines resistant to strains that rely on *PthXo1*. This finding also demonstrated that *Os-8N3* is required for susceptibility.

Xa13/Os-8N3 is required for pollen development, showing that such mutant alleles can be problematic should the disease-susceptibility phenotype alter function in other processes. However, mutations in the *Os11N3* (OsSWEET14) TAL effector–binding element were made by fusing TAL effectors to nucleases (TALENs). Genome-edited rice plants with altered *Os11N3* binding sites remained resistant to *Xanthomonas oryzae pv. oryzae*, but still provided normal development function.

## Gene Silencing

RNA silencing-based resistance is a powerful tool for engineering resistant crops. The advantage of RNAi as a novel gene therapy against fungal, viral and bacterial infection in plants lies in the fact that it regulates gene expression via messenger RNA degradation, translation repression and chromatin remodelling through small non-coding RNAs. Mechanistically, the silencing processes are guided by processing products of the double-stranded RNA (dsRNA) trigger, which are known as small interfering RNAs and microRNAs.

## Host Range

Among the thousands of species of plant pathogenic microorganisms, only a small minority have the capacity to infect a broad range of plant species. Most pathogens instead exhibit a high degree of host-specificity. Non-host plant species are often said to express *non-host resistance*. The term *host resistance* is used when a pathogen species can be pathogenic on the host species but certain strains of that plant species resist certain strains of the pathogen species. The causes of host resistance and non-host resistance can overlap. Pathogen host range can change quite suddenly if, for example, the pathogen's capacity to synthesize a host-specific toxin or effector is gained by gene shuffling/mutation, or by horizontal gene transfer.

## Epidemics and Population Biology

Native populations are often characterized by substantial genotype diversity and dispersed populations (growth in a mixture with many other plant species). They also have undergone of plant-pathogen coevolution. Hence as long as novel pathogens are not introduced/do not evolve, such populations generally exhibit only a low incidence of severe disease epidemics.

Monocrop agricultural systems provide an ideal environment for pathogen evolution, because they offer a high density of target specimens with similar/identical genotypes.

The rise in mobility stemming from modern transportation systems provides pathogens with access to more potential targets.

Climate change can alter the viable geographic range of pathogen species and cause some diseases to become a problem in areas where the disease was previously less important.

These factors make modern agriculture more prone to disease epidemics. Common solutions include constant breeding for disease resistance, use of pesticides, use of border inspections and plant import restrictions, maintenance of significant genetic diversity within the crop gene pool and constant surveillance to accelerate initiation of appropriate responses. Some pathogen species have much greater capacity to overcome plant disease resistance than others, often because of their ability to evolve rapidly and to disperse broadly.

# Disease Resistance in Fruit and Vegetables

There are a number of lines of defence against pests (that, those animals that cause damage to the plants we grow) and diseases in the O, principal among these being the practice of good husbandry, creating healthy soil and ensuring high standards of garden hygiene. But no matter how diverse and healthy the garden eco-system may be, there will always be a degree of disease and pest presence. In many ways, some level of pathogen population in the garden can be not only acceptable but desirable as they are indicative of a generally healthful and diverse environment, and add to the overall robustness of the system as an immunity to such detrimental influences will build up, particularly in a balanced polycultural regime. Indeed, most of the plants we grow will tend to be selected because they are trouble free, and those that are more susceptible to attack will have fallen by the wayside over time. However, most farmers find it unacceptable that the food crops they grow are damaged by pests.

For these crops there has been considerable research and selective breeding carried out in order to find cultivars that are resistant or immune to pest and disease damage. Breeding for plant disease resistance generally has involved finding suitable genetic material amongst existing stocks or in the wild, which is then incorporated into commercial varieties.

### Example: the Apple

In the case of apples, in which research is being carried out in order to develop resistance to diseases such as apple scab (*Venturia inaequalis*), powdery mildew (*Podosphaera leucotricha*), orchard fireblight (*Erwinia amylovora*), woolly apple aphid (*Eriosoma lanigerum*) and collar rot (*Phytophthora cactorum*), the main sources of resistant material used in breeding programmes such as those being run by East Malling Research Station in England or Hortresearch in New Zealand are major gene resistances derived from crab-apples. The Vf gene for black spot resistance is derived from the ornamental crab-apple species *Malus floribunda*. Most black spot resistant cultivars developed around the world carry this gene, but there are some selections that carry the Vr (from M. pumila) or Vm (from M. micromalus) gene. Major gene resistances to powdery mildew are derived from M. robusta (Pl1) and M. zumi (Pl2), and the apple cultivar *Northern Spy* has a long-standing reputation for its major gene resistance to woolly apple aphid. Since early this century this resistance has been used to develop woolly aphid resistant rootstocks such as MM.106 and M.793. Much later it was shown that the cultivar was also very resistant to collar rot and a useful breeding parent for this resistance.

## Resistance and Immunity

Some plants can tolerate the presence of large numbers of insects without being severely affected. This is not very satisfactory however as insects will still cause damage, and in fact further breeding and population expansion of the pest species is supported. Other varieties are less attractive to pests, but this can be difficult to sustain or demonstrate. The most valuable form of resistance is where the pest cannot survive as well on one variety as on another. In some cases this can actually make the plants immune to attack, as is the case with the lettuces *Avoncrisp* and *Avondefiance* which were bred at the Institute of Horticultural Research, Wellesbourne during the 1960s, which are fully resistant to lettuce root aphid (*Pemphigus bursarius*).

## Trade-off of Breeding for Resistance

Sometimes however there can be a *trade-off*, for those varieties which have increased immunity or resistance may be lacking in other qualities such as flavour, yield or quality. Celery resistant to the Fusarium fungus (*Fusarium oxysporum spp.*) may not succumb to this disease, but may also be unacceptably short, ribby and low yielding. Further, a cultivar that is resistant to one disease may be more susceptible to another that is equally important. A lettuce cultivar that is resistant to mosaic virus may be sensitive to corky root disease, whilst another that resists corky root may be vulnerable to downy mildew (*Brim lactic*). Another drawback to resistance is that depending on the host pathogen system, resistance is sometimes not long lasting as new pathogen strains quickly develop, and further research and breeding is constantly needed.

## Availability of Resistant Varieties

Resistant varieties are not available for all crops. For several of the most damaging plant diseases, such as Potato blight (*Phytophthora infestans*) and white rot (*Sclerotic cepivorum*) of the Allium family, no acceptable resistant cultivars are yet available. In addition, commercial seed companies and plant breeders rarely invest resources into developing resistant cultivars for more minor or speciality crops, which often tend to be those of greater interest to the organic grower.

In general it is probably fair to say that resistance will not fully guarantee total crop protection, but choosing resistant varieties should rather be considered as a part of an overall Integrated pest management strategy, especially against virus diseases. In particular they can be especially useful where the threat from specific pests and diseases is high.

# Pest Control

Pest control refers to the regulation or management of a species defined as a pest, and can be perceived to be detrimental to a person's health, the ecology or the economy. A practitioner of pest control is called an exterminator.

A crop duster applies low-insecticide bait that is targeted against western corn rootworms.

## History

Pest control is at least as old as agriculture, as there has always been a need to keep crops free from pests. In order to maximize food production, it is advantageous to protect crops from competing species of plants, as well as from herbivores competing with humans.

The conventional approach was probably the first to be employed, since it is comparatively easy to destroy weeds by burning them or plowing them under, and to kill larger competing herbivores, such as crows and other birds eating seeds. Techniques such as crop rotation, companion planting (also known as intercropping or mixed cropping), and the selective breeding of pest-resistant cultivars have a long history.

In the UK, following concern about animal welfare, humane pest control and deterrence is gaining ground through the use of animal psychology rather than destruction. For instance, with the urban red fox which territorial behaviour is used against the animal, usually in conjunction with non-injurious chemical repellents. In rural areas of Britain, the use of firearms for pest control is quite common. Airguns are particularly popular for control of small pests such as rats, rabbits and grey squirrels, because of their lower power they can be used in more restrictive spaces such as gardens, where using a firearm would be unsafe.

Chemical pesticides date back 4,500 years, when the Sumerians used sulfur compounds as insecticides. The Rig Veda, which is about 4,000 years old, also mentions the use of poisonous plants for pest control. It was only with the industrialization and mechanization of agriculture in the 18th and 19th century, and the introduction of the insecticides pyrethrum and derris that chemical pest control became widespread. In the 20th century, the discovery of several synthetic insecticides, such as DDT, and herbicides boosted this development. Chemical pest control is still the predominant type of pest control today, although its long-term effects led to a renewed interest in traditional and biological pest control towards the end of the 20th century.

## Causes

Sign in Ilfracombe, England designed to help control seagull presence

Many pests have only become a problem as a result of the direct actions by humans. Modifying these actions can often substantially reduce the pest problem. In the United States, raccoons caused a nuisance by tearing open refuse sacks. Many householders introduced bins with locking lids, which deterred the raccoons from visiting. House flies tend to accumulate wherever there is human activity and is virtually a global phenomenon, especially where food or food waste is exposed. Similarly, seagulls have become pests at many seaside resorts. Tourists would often feed the birds with scraps of fish and chips, and before long, the birds would rely on this food source and act aggressively towards humans.

Living organisms evolve and increase their resistance to biological, chemical, physical or any other form of control. Unless the target population is completely exterminated or is rendered incapable of reproduction, the surviving population will inevitably acquire a tolerance of whatever pressures are brought to bear - this results in an evolutionary arms race.

## Types of Pest Control

## Use of Pest-destroying Animals

Perhaps as far ago as 3000BC in Egypt, cats were being used to control pests of grain stores such as rodents. In 1939/40 a survey discovered that cats could keep a farm's population of rats down to a low level, but could not eliminate them completely. However, if the rats were cleared by trapping or poisoning, farm cats could stop them returning - at least from an area of 50 yards around a barn.

Ferrets were domesticated at least by 500 AD in Europe, being used as mousers. Mongooses have been introduced into homes to control rodents and snakes, probably at first by the ancient Egyptians.

## Biological Pest Control

Biological pest control is the control of one through the control and management of natural predators and parasites. For example: mosquitoes are often controlled by putting *Bt Bacillus thuringiensis* ssp. *israelensis*, a bacterium that infects and kills mosquito larvae, in local water sources. The treatment has no known negative consequences on the remaining ecology and is safe for humans to drink. The point of biological pest control, or any natural pest control, is to eliminate a pest with minimal harm to the ecological balance of the environment in its present form.

## Mechanical Pest Control

Mechanical pest control is the use of hands-on techniques as well as simple equipment and devices, that provides a protective barrier between plants and insects. For example: weeds can be controlled by being physically removed from the ground. This is referred to as tillage and is one of the oldest methods of weed control.

## Physical Pest Control

Dog control van, Rekong Peo, Himachal Pradesh, India

Physical pest control is a method of getting rid of insects and small rodents by removing, attacking, setting up barriers that will prevent further destruction of one's plants, or forcing insect infestations to become visual.

## Elimination of Breeding Grounds

Proper waste management and drainage of still water, eliminates the breeding ground of many pests.

Garbage provides food and shelter for many unwanted organisms, as well as an area where still water might collect and be used as a breeding ground by mosquitoes. Communities that have proper garbage collection and disposal, have far less of a problem with rats, cockroaches, mosquitoes, flies and other pests than those that don't.

Open air sewers are ample breeding ground for various pests as well. By building and maintaining a proper sewer system, this problem is eliminated.

Certain spectrums of LED light can "disrupt insects' breeding".

## Poisoned Bait

Poisoned bait is a common method for controlling rat populations, however is not as effective when there are other food sources around, such as garbage. Poisoned meats have been used for centuries for killing off wolves, birds that were seen to threaten crops, and against other creatures. This can be a problem, since a carcass which has been poisoned will kill not only the targeted animal, but also every other animal which feeds on the carcass. Humans have also been killed by coming in contact with poisoned meat, or by eating an animal which had fed on a poisoned carcass. This tool is also used to manage several caterpillars e.g. Spodoptera litura, fruit flies, snails and slugs, crabs etc.

## Field Burning

Traditionally, after a sugar cane harvest, the fields are all burned, to kill off any rodents, insects or eggs that might be in the fields.

## Hunting

Historically, in some European countries, when stray dogs and cats became too numerous, local populations gathered together to round up all animals that did not appear to have an owner and kill them. In some nations, teams of rat-catchers work at chasing rats from the field, and killing them with dogs and simple hand tools. Some communities have in the past employed a bounty system, where a town clerk will pay a set fee for every rat head brought in as proof of a rat killing.

## Traps

A variety of mouse traps and rat traps are available for mice and rats, including snap traps, glue traps and live catch traps.

## Pesticides

Spraying pesticides by planes, trucks or by hand is a common method of pest control. Crop dusters commonly fly over farmland and spray pesticides to kill off pests that would threaten the crops. However, some pesticides may cause cancer and other health problems, as well as harming wild-life.

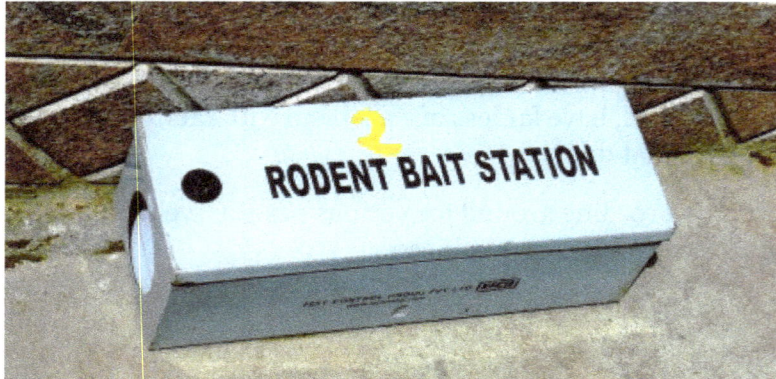

Rodent bait station, Chennai, India

## Space Fumigation

A project that involves a structure be covered or sealed airtight followed by the introduction of a penetrating, deadly gas at a killing concentration a long period of time (24-72hrs.). Although expensive, space fumigation targets all life stages of pests.

## Space Treatment

Residential & commercial building pest control service vehicle, Ypsilanti Township, Michigan

A long term project involving fogging or misting type applicators. Liquid insecticide is dispersed in the atmosphere within a structure. Treatments do not require the evacuation or airtight sealing of a building, allowing most work within the building to continue but at the cost of the penetrating effects. Contact insecticides are generally used, minimizing the long lasting residual effects. On August 10, 1973, the Federal Register printed the definition of Space treatment as defined by the U.S. Environmental Protection Agency (EPA):

> "The dispersal of insecticides into the air by foggers, misters, aerosol devices or vapor dispensers for control of flying insects and exposed crawling insects"

## Sterilization

Laboratory studies conducted with U-5897 (3-chloro-1,2-propanediol) were attempted in the early 1970s although these proved unsuccessful. Research into sterilization bait is ongoing.

In 2013, New York City tested sterilization traps in a $1.1 million study. The result was a 43% reduction in rat populations. The Chicago Transit Authority plans to test sterilization control in spring 2015. The sterilization method doesn't poison the rats or humans.

## Destruction of Infected Plants

Forest services sometimes destroy all the trees in an area where some are infected with insects, if seen as necessary to prevent the insect species from spreading. Farms infested with certain insects, have been burned entirely, to prevent the pest from spreading elsewhere.

## Natural Rodent Control

Example of House mouse infestation

Several wildlife rehabilitation organizations encourage natural form of rodent control through exclusion and predator support and preventing secondary poisoning altogether.

The United States Environmental Protection Agency agrees, noting in its Proposed Risk Mitigation Decision for Nine Rodenticides that "without habitat modification to make areas less attractive to commensal rodents, even eradication will not prevent new populations from recolonizing the habitat."

## Repellents

- Balsam fir oil from the tree *Abies balsamea* is an EPA approved non-toxic rodent repellent.
- *Acacia polyacantha* subsp. *campylacantha* root emits chemical compounds that repel animals including crocodiles, snakes and rats.

# Fungicide

Fungicides are biocidal chemical compounds or biological organisms used to kill fungi or fungal spores. A fungistatic inhibits their growth. Fungi can cause serious damage in agriculture, resulting in critical losses of yield, quality, and profit. Fungicides are used both in agriculture and to fight fungal infections in animals. Chemicals used to control oomycetes, which are not fungi, are also referred to as fungicides, as oomycetes use the same mechanisms as fungi to infect plants.

Fungicides can either be contact, translaminar or systemic. Contact fungicides are not taken up into the plant tissue and protect only the plant where the spray is deposited. Translaminar fungicides redistribute the fungicide from the upper, sprayed leaf surface to the lower, unsprayed surface. Systemic fungicides are taken up and redistributed through the xylem vessels. Few fungicides move to all parts of a plant. Some are locally systemic, and some move upwardly.

Most fungicides that can be bought retail are sold in a liquid form. A very common active ingredient is sulfur, present at 0.08% in weaker concentrates, and as high as 0.5% for more potent fungicides. Fungicides in powdered form are usually around 90% sulfur and are very toxic. Other active ingredients in fungicides include neem oil, rosemary oil, jojoba oil, the bacterium *Bacillus subtilis*, and the beneficial fungus *Ulocladium oudemansii*.

Fungicide residues have been found on food for human consumption, mostly from post-harvest treatments. Some fungicides are dangerous to human health, such as vinclozolin, which has now been removed from use. Ziram is also a fungicide that is thought to be toxic to humans if exposed to chronically. A number of fungicides are also used in human health care.

## Natural Fungicides

Plants and other organisms have chemical defenses that give them an advantage against microorganisms such as fungi. Some of these compounds can be used as fungicides:

- Tea tree oil
- Cinnamaldehyde
- Citronella oil
- Jojoba oil
- Nimbin
- Oregano oil
- Rosemary oil
- Monocerin
- Milk

Whole live or dead organisms that are efficient at killing or inhibiting fungi can sometimes be used as fungicides:

- *Bacillus subtilis*
- *Ulocladium oudemansii*

- Kelp (powdered dried kelp is fed to cattle to help prevent fungal infection)

- *Ampelomyces quisqualis*

## Resistance

Pathogens respond to the use of fungicides by evolving resistance. In the field several mechanisms of resistance have been identified. The evolution of fungicide resistance can be gradual or sudden. In qualitative or discrete resistance, a mutation (normally to a single gene) produces a race of a fungus with a high degree of resistance. Such resistant varieties also tend to show stability, persisting after the fungicide has been removed from the market. For example, sugar beet leaf blotch remains resistant to azoles years after they were no longer used for control of the disease. This is because such mutations often have a high selection pressure when the fungicide is used, but there is low selection pressure to remove them in the absence of the fungicide.

In instances where resistance occurs more gradually, a shift in sensitivity in the pathogen to the fungicide can be seen. Such resistance is polygenic – an accumulation of many mutations in different genes, each having a small additive effect. This type of resistance is known as quantitative or continuous resistance. In this kind of resistance, the pathogen population will revert to a sensitive state if the fungicide is no longer applied.

Little is known about how variations in fungicide treatment affect the selection pressure to evolve resistance to that fungicide. Evidence shows that the doses that provide the most control of the disease also provide the largest selection pressure to acquire resistance, and that lower doses decrease the selection pressure.

In some cases when a pathogen evolves resistance to one fungicide, it automatically obtains resistance to others – a phenomenon known as cross resistance. These additional fungicides are normally of the same chemical family or have the same mode of action, or can be detoxified by the same mechanism. Sometimes negative cross resistance occurs, where resistance to one chemical class of fungicides leads to an increase in sensitivity to a different chemical class of fungicides. This has been seen with carbendazim and diethofencarb.

There are also recorded incidences of the evolution of multiple drug resistance by pathogens – resistance to two chemically different fungicides by separate mutation events. For example, *Botrytis cinerea* is resistant to both azoles and dicarboximide fungicides.

There are several routes by which pathogens can evolve fungicide resistance. The most common mechanism appears to be alteration of the target site, in particular as a defence against single site of action fungicides. For example, Black Sigatoka, an economically important pathogen of banana, is resistant to the QoI fungicides, due to a single nucleotide change resulting in the replacement of one amino acid (glycine) by another (alanine) in the target protein of the QoI fungicides, cytochrome b. It is presumed that this disrupts the binding of the fungicide to the protein, rendering the fungicide ineffective. Upregulation of target genes can also render the fungicide ineffective. This is seen in DMI-resistant strains of *Venturia inaequalis*.

Resistance to fungicides can also be developed by efficient efflux of the fungicide out of the cell. *Septoria tritici* has developed multiple drug resistance using this mechanism. The pathogen had

5 ABC-type transporters with overlapping substrate specificities that together work to pump toxic chemicals out of the cell.

In addition to the mechanisms outlined above, fungi may also develop metabolic pathways that circumvent the target protein, or acquire enzymes that enable metabolism of the fungicide to a harmless substance.

## Fungicide Resistance Management

The fungicide resistance action committee (FRAC) has several recommended practices to try to avoid the development of fungicide resistance, especially in at-risk fungicides including *Strobilurins* such as azoxystrobin.

Products should not be used in isolation, but rather as mixture, or alternate sprays, with another fungicide with a different mechanism of action. The likelihood of the pathogen's developing resistance is greatly decreased by the fact that any resistant isolates to one fungicide will be killed by the other; in other words, two mutations would be required rather than just one. The effectiveness of this technique can be demonstrated by Metalaxyl, a phenylamide fungicide. When used as the sole product in Ireland to control potato blight (*Phytophthora infestans*), resistance developed within one growing season. However, in countries like the UK where it was marketed only as a mixture, resistance problems developed more slowly.

Fungicides should be applied only when absolutely necessary, especially if they are in an at-risk group. Lowering the amount of fungicide in the environment lowers the selection pressure for resistance to develop.

Manufacturers' doses should always be followed. These doses are normally designed to give the right balance between controlling the disease and limiting the risk of resistance development. Higher doses increase the selection pressure for single-site mutations that confer resistance, as all strains but those that carry the mutation will be eliminated, and thus the resistant strain will propagate. Lower doses greatly increase the risk of polygenic resistance, as strains that are slightly less sensitive to the fungicide may survive.

It is also recommended that where possible fungicides are used only in a protective manner, rather than to try to cure already-infected crops. Far fewer fungicides have curative/eradicative ability than protectant. Thus, fungicide preparations advertised as having curative action may have only one active chemical; a single fungicide acting in isolation increases the risk of fungicide resistance.

It is better to use an integrative pest management approach to disease control rather than relying on fungicides alone. This involves the use of resistant varieties and hygienic practices, such as the removal of potato discard piles and stubble on which the pathogen can overwinter, greatly reducing the titre of the pathogen and thus the risk of fungicide resistance development.

## References

- Fred Baur. Insect Management for Food Storage and Processing. American Association of Cereal Chemists. ISBN 0-913250-38-4.

- "Help WildCare Pursue Stricter Rodenticide Controls in California". wildcarebayarea.org/. Wild Care. Retrieved 28 February 2014.

- "Safer Rodenticide Products". epa.gov. USA Environment Protection Agency. March 2013. Retrieved 23 February 2014.

- WOODY, TODD (September 20, 2010). "A Crop Sprouts Without Soil or Sunshine". nytimes.com. The New York Times. Retrieved 28 February 2014.

- "Pesticides". National Institute of Health Sciences. National Institute of Environmental Health. Retrieved 5 April 2013.

# Defense Mechanisms of Plants

Thigmonasty is the response of a plant or fungus to touch or vibration, while the ability of reducing the negative fitness effects caused by herbivory is known as the plant tolerance to herbivory. This chapter elaborates on other defense mechanisms also, which include, raphide, inducible plan defenses against herbivory and plant use of endophytic fungi in defense.

## Thigmonasty

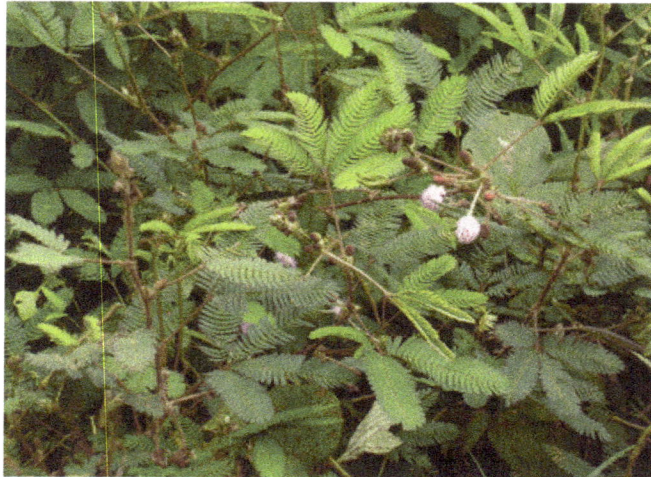

*Mimosa pudica* in normal and touched state.

Thigmonasty or seismonasty is the nastic response of a plant or fungus to touch or vibration. Conspicuous examples of thigmonasty include many species in the leguminous subfamily Mimosoideae, active carnivorous plants such as Dionaea and a wide range of pollination mechanisms.

### Distinctive Aspects of Thigmonasty

Thigmonasty differs from thigmotropism in that nastic motion is independent of the direction of the stimulus. For example, tendrils from a climbing plant are thigmotropic because they twine around any support they touch, responding in whichever direction the stimulus came from. However, the shutting of a venus fly trap is thigmonastic; no matter what the direction of the stimulus, the trap simply shuts (and later possibly opens).

The time scales of thigmonastic responses tend to be shorter than those of thigmotropic movements because many examples of thigmonasty depend on pre-accumulated turgor or on bistable mechanisms rather than growth or cell division. Certain dramatic examples of rapid plant movement such as the sudden drooping of *Mimosa pudica* or the trapping action of *Dionaea* or *Utric-*

*ularia* are fast enough to observe without time lapse photography; some take less than a second. Speed is no clear distinction however; for example the re-erection of *Mimosa* leaves is nastic, but typically takes some 15 to 30 minutes, rather than a second or so. Similarly, re-opening of the *Dionaea* trap, though also nastic, typically takes days to complete.

Botanical physiologists have discovered signalling molecules called turgorins, that help mediate the loss of turgor. In species with the fastest response time, vacuoles are believed to provide temporary, high speed storage for calcium ions.

## Examples of Plants Exhibiting Thigmonasty

### Thigmonasty in the Asteraceae

Thigmonasty other than leaf closure occurs in various species of thistles. When an insect lands on a flower, the anthers shrink and rebound, loading the insect with pollen. The effect results from turgor changes in specialized, highly elastic cell walls of the anthers. Similar pollination strategy occurs in *Rudbeckia hirta*.

### Thigmonasty in the Droseraceae

The Venus Flytrap (*Dionaea muscipula*) presents a spectacular example of thigmonasty; when an insect lands on a trap formed by two curved lobes of a single leaf, the trap rapidly switches from an open to a closed configuration. Investigators have observed an action potential and changes in leaf turgor that accompany the reflex; they trigger the rapid elongation of individual cells. The common term for the elongation is acid growth although the process does not involve cell division.

### Thigmonasty in the Fabaceae

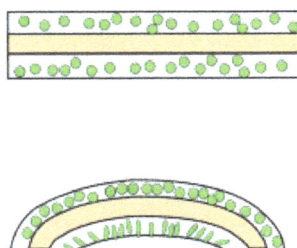

Pulvinus in extended and contracted position

*Mimosa pudica* is a plant with compound leaves that droop abruptly when stimulated. This is a classic example of thigmonastic action and has attracted detailed investigation. Contact or injury that causes leaflets to deform, will trigger an action potential. The action potential travels through the plant, initiating drooping of the leaflets as it passes. However, it cannot pass the pulvinus at the base of a petiole, and this prevents a local disturbance from causing all the leaves on the plant to collapse in response to trivial disturbances.

The pulvinus is a motor structure consisting of a rod of sclerenchyma surrounded by collenchyma. Such pulvini occur widely in the Fabaceae. In its extended position, the cells of the entire collar of collenchyma are distended with water. On receiving the action potential signal, the cells in the lower half of the pulvinus respond by expelling potassium and chlorine ions and taking up of calcium ions. This results in an osmotic gradient that draws water out of the affected cells, so that they temporarily shrink. This pulls the entire structure downward like a folding fan.

Many other Fabaceae react to touch with the same rapid leaf closure motion. These include the telegraph plant and the silk tree. The pea vine thigmonastically closes its leaves around a support. Catclaw Brier, a prairie mimosa, native to North America, shuts its leaves on contact. The plant is attractive to herbivores, and this behavior presumably provides protection against grazing.

### Thigmonasty in The Oxalidaceae

Sensitive leaves also occur in plants of the wood sorrel family. Examples include many species of *Oxalis*, *Biophytum sensitivum*, and Averrhoa carambola (the plant which produces starfruit).

### Other Forms

Some fungi exhibit trap closure similar to the venus fly trap. Mycologists have discovered action potentials in fungi but it is not currently clear whether they have any significance to thigmonastic behavior.

## Raphide

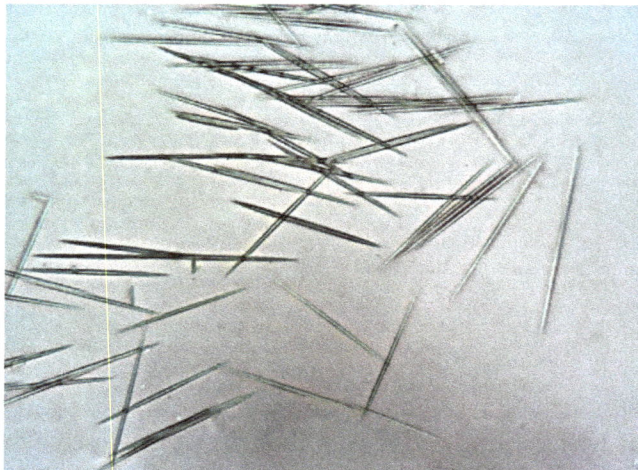

Raphides in *Epipremnum* Devil's ivy (600x magnification)

Raphides are needle-shaped crystals of calcium oxalate as the monohydrate or calcium carbonate as aragonite, found in more than 200 families of plants. Both ends are needle-like, but raphides tend to be blunt at one end and sharp at the other.

## Calcium Oxalate in Plants

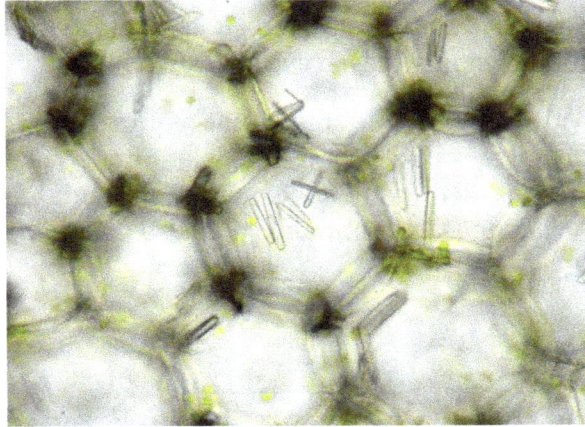

Raphides in *Hypoestes phyllostachya*, the polka dot plant

Many plants accumulate calcium oxalate crystals in response to surplus calcium, which is found throughout the natural environment. The crystals are produced in an intriguing variety of shapes. The crystal morphology depends on the taxonomic group of the plant. In one study of over 100 species, it was found that calcium oxalate accounted for 6.3% of plant dry weight. Crystal morphology and the distribution of raphides (in roots or leaves or tubers etc.) is similar in same taxa but different in others leaving possible opportunities for plant key characteristics and systematic identification; mucilage in raphide containing cells makes light microscopy difficult, though. Little is known about the mechanisms of sequestration or indeed the reason for accumulation of raphides but it is most likely as a defense mechanism against herbivory. It has also been suggested that in some cases raphides may help form plant skeletal structure. Raphides typically occur in parenchyma cells in aerial organs especially the leaves, and are generally confined to the mesophyl. As the leaf area increases, so does the number of raphides, the process starting in even young leaves. The first indications that the cell will contain crystals is shown when the cells enlarge with a larger nucleus.

Raphides are found in specialized plant cells or crystal chambers called idioblasts. Electron micrographs have shown that raphide needle crystals are normally four sided or H-shaped (with a groove down both sides) or with a hexagonal cross section and some are barbed. Wattendorf (1976) suggested that all circular sectioned raphides, as visible in a light microscope, are probably hexagonal in cross section Microscopy using polarized light shows bright opalescence with raphides. Plants like Tradescantia pallida also accumulate calcium oxalate crystals in response to heavy metals stress.

## Harmful Effects

Raphides can produce severe toxic reactions by facilitating the passage of toxin through the herbivore's skin when the tissue containing the raphides also contains toxins. The lethal dose to mice is around 15 mg/kg. Raphides seem to be a defense mechanism against plant predators, as they are likely to tear and harm the soft tissues of the throat or esophagus of a plant predator chewing on the plant's leaves. The venomous process is in two stages: mechanical pricking and injection of harmful protease. Typically ingestion of plants containing raphides, like those common in certain houseplants, can cause immediate numbing followed shortly by painful edema, vesicle formation and dysphagia accompanied by painful stinging and burning to the mouth and throat with symptoms occurring for up to two weeks. Airway assessment and management are of the highest priority, as are extensive irrigation and analgesics in eye exposure.

Raphides cannot normally be destroyed by boiling; that requires an acidic environment or chemical solvents like ether, but heating raphide containing plant materials (like tubers) can fix the raphides into a dried starchy matrix so they are no longer mobile and thus less harmful. Some other plants store raphides in mucilaginous environments and also do not taste acrid.

# Plant Tolerance to Herbivory

Tolerance is the ability of plants to mitigate the negative fitness effects caused by herbivory. It is one of the general plant defense strategies against herbivores, the other being resistance, which is the ability of plants to prevent damage (Strauss and Agrawal 1999). Plant defense strategies play important roles in the survival of plants as they are fed upon by many different types of herbivores, especially insects, which may impose negative fitness effects (Strauss and Zangerl 2002). Damage can occur in almost any part of the plants, including the roots, stems, leaves, flowers and seeds (Strauss and Zergerl 2002). In response to herbivory, plants have evolved a wide variety of defense mechanisms and although relatively less studied than resistance strategies, tolerance traits play a major role in plant defense (Strauss and Zergerl 2002, Rosenthal and Kotanen 1995).

Traits that confer tolerance are controlled genetically and therefore are heritable traits under selection (Strauss and Agrawal 1999). Many factors intrinsic to the plants, such as growth rate, storage capacity, photosynthetic rates and nutrient allocation and uptake, can affect the extent to which plants can tolerate damage (Rosenthal and Kotanen 1994). Extrinsic factors such as soil nutrition, carbon dioxide levels, light levels, water availability and competition also have an effect on tolerance (Rosenthal and Kotanen 1994).

### History of The Study of Plant Tolerance

Studies of tolerance to herbivory has historically been the focus of agricultural scientists (Painter 1958; Bardner and Fletcher 1974). Tolerance was actually initially classified as a form of resistance (Painter 1958). Agricultural studies on tolerance, however, are mainly concerned with the compensatory effect on the plants' yield and not its fitness, since it is of economical interest to reduce crop losses due to herbivory by pests (Trumble 1993; Bardner and Fletcher 1974). One surprising discovery made about plant tolerance was that plants can overcompensate for the damaged caused by herbivory, causing controversy whether herbivores and plants can actually form a mutualistic relationship (Belsky 1986).

It was soon recognized that many factors involved in plants tolerance, such as photosynthetic rates and nutrient allocation, were also traits intrinsic to plant growth and so resource availability may play an important role (Hilbert et al. 1981; Maschinski and Whitham 1989). The growth rate model proposed by Hilbert et al. (1981) predicts plants have higher tolerance in environments that does not allow it to grow at maximum capacity, while the compensatory continuum hypothesis by Maschinski and Whitham (1989) predicts higher tolerance in resource rich environments. Although it was the latter that received higher acceptance, 20 years later, the limiting resource model was proposed to explain the lack of agreement between empirical data and existing models (Wise and Abrahamson 2007). Currently, the limiting resource model is able to explain much more of the empirical data on plant tolerance relative to either of the previous models (Wise and Abrahamson 2008a).

It was only recently that the assumption that tolerance and resistance must be negatively associated has been rejected (Nunez-Farfan et al. 2007). The classical assumption that tolerance traits confer no negative fitness consequences on herbivores has also been questioned (Stinchcombe 2002). Further studies using techniques in quantitative genetics have also provided evidence that tolerance to herbivory is heritable (Fornoni 2011). Studies of plant tolerance have only received increased attention recently, unlike resistance traits which were much more heavily studied (Fornoni 2011). Many aspects of plant tolerance such as its geographic variation, its macroevolutionary implications and its coevolutionary effects on herbivores are still relatively unknown (Fornoni 2011).

## Mechanisms of Tolerance

Plants utilize many mechanisms to recover fitness from damage. Such traits include increased photosynthetic activity, compensatory growth, phenological changes, utilizing stored reserves, reallocating resources, increase in nutrients uptake, and plant architecture (Rosenthal and Kotanen 1994; Strauss and Agrawal 1999; Tiffin 2000).

## Photosynthetic Rates

An increase in photosynthetic rate in undamaged tissues is commonly cited as a mechanism for plants to achieve tolerance (Trumble et al. 1993; Strauss and Agrawal 1999). This is possible since leaves often function at below their maximum capacity (Trumble et al. 1993). Several different pathways may lead to increases in photosynthesis, including higher levels of the Rubisco enzyme and delays in leaf senescence (Stowe et al. 2000). However, detecting an increase in photosynthesis does not mean plants are tolerant to damage. The resources gained from these mechanisms can be used to increase resistance instead of tolerance, such as for the production secondary compounds in the plant (Tiffin 2000). Also, whether the increase in photosynthetic rate is able to compensate for the damage is still not well studied (Trumble et al. 1993; Stowe et al. 2000).

## Compensatory Growth

Biomass regrowth following herbivory is often reported as an indicator of tolerance and plant response after apical meristem damage (AMD) is one of the most heavily studied mechanisms of tolerance (Tiffin 2000; Suwa and Maherali 2008; Wise and Abrahamson 2008). Meristems are sites of rapid cell divisions and so have higher nutrition than most other tissues on the plants . Damage to apical meristems of plants may release it from apical dominance, activating the growth of axillary meristems which increases branching (Trumble et al. 1993; Wise and Abrahamson 2008). Studies have found branching after AMD to undercompensate, fully compensate and overcompensate for the damage received (Marquis 1996, Haukioja and Koricheva 2000, Wise and Abrahamson 2008). The variation in the extent of growth following herbivory may depend on the number and distribution of meristems, the pattern in which they are activated and the number of new meristems

(Stowe et al. 2000). The wide occurrence of overcompensation after AMD has also brought up a controversial idea that there may be a mutualistic relationship between plants and their herbivores (Belsky 1986; Agrawal 2000; Edwards 2009). As will be further discussed below, herbivores may actually be mutalists of plants, such as *Ipomopsis aggregata*, which overcompensate for herbivory (Edwards 2009). Although there are many examples showing biomass regrowth following herbivory, it has been criticized as a useful predictor of fitness since the resources used for regrowth may translate to fewer resources allocated to reproduction (Suwa and Maherali 2008).

## Phenological Change

Studies have shown herbivory can cause delays in plant growth, flowering and fruit production (Tiffin 2000). How plants respond to these phenological delays is likely a tolerance mechanism that will depend highly on their life history and other ecological factors such as, the abundance of pollinators at different times during the season (Tiffin 2000). If the growing season is short, plants that are able to shorten the delay of seed production caused by herbivory are more tolerant than those that cannot shorten this phenological change (Tiffin 2000). These faster recovering plant will be selectively favored over those that cannot as they will pass on more of their offspring to the next generation. In longer growing seasons, however, there may be enough time for most plants to produce seeds before the season ends regardless of damage. In this case, plants that can shorten the phenological delay are not any more tolerant than those that cannot as all plants can reproduce before the season ends (Tiffin 2000).

## Stored Reserves and Resource Reallocation

Resource allocation following herbivory is commonly studied in agricultural systems (Trumble et al. 1993). Resources are most often allocated to reproductive structures after damage, as shown by Irwin et al. (2008) in which *Polemonium viscosum* and *Ipomopsis aggregata* increased flower production after flower larceny. When these reproductive structures are not present, resources are allocated to other tissues, such as leaves and shoots as seen in juvenile *Plantago lanceolata* (Trumble et al. 1993; Barton 2008). Utilizing stored reserves may be an important tolerance mechanism for plants which have abundant time to collect and store resources, such as perennial plants (Tiffin 2000; Erb *et al.* 2009). Resources are often stored in leaves and specialized storage organs such as tubers and roots, and studies have shown evidence that these resources are allocated for regrowth following herbivory (Trumble et al. 1993; Tiffin 2000; Erb *et al.* 2009). However, the importance of this mechanism to tolerance is not well studied and it is unknown how much it contributes to tolerance since stored reserves mostly consist of carbon resources, whereas tissue damage causes a loss of carbon, nitrogen and other nutrients (Tiffin 2000).

## Plant Architecture

This form of tolerance relies on constitutive mechanisms, such as morphology, at the time of damage, unlike the induced mechanisms mentioned above. Plant architecture includes roots to shoots ratios, stem number, stem rigidity and plant vasculature (Marquis 1996, Tiffin 2000). A high roots to shoots ratio will allow plants to better absorb nutrients following herbivory and rigid stems will prevent collapse after sustaining damage, increasing plant tolerance (Tiffin 2000). Since plants have a meristemic construction, how resources are restricted among different regions of the plants, referred to as sectoriality, will affect the ability to transfer resources from undamaged areas to damaged areas (Marquis

1996). Although plant vasculature may play important roles in tolerance, it is not well studied due to the difficulties in identifying the flow of resources (Marquis 1996). Increasing a plant's vasculature would seem advantageous since it increases the flow of resources to all sites of damage but it may also increase its susceptibility to herbivores, such as phloem suckers (Marquis 1996, Stowe et al. 2000).

## Measuring Tolerance

Tolerance is operationally defined as the slope of the regression between fitness and level of damage (Stinchcombe 2002). Since an individual plant can only sustain one level of damage, it is necessary to measure fitness using a group of related individuals, preferably full-sibs or clones to minimize other factors that may influence tolerance, after sustaining different levels of damage (Stinchcombe 2002). Tolerance is often presented as a reaction norm, where slopes larger than, equal to and less than zero reflect overcompensation, full compensation and undercompensation, respectively (Strauss and Agrawal 1999).

## Scales of Measurement

Both fitness and herbivory can be measured or analyzed using an absolute (additive) scale or a relative (multiplicative) scale (Wise and Carr 2008b). The absolute scale may refer to number of fruits produced or total area of leaf eaten, while the relative scale may refer to proportion of fruits damaged or proportion of leaves eaten. Wise and Carr (2008b) suggested that it is best to keep the measure of fitness and the measure of damage on the same scale when analyzing tolerance since having them on different scales may result is misleading outcomes. Even if the data were measured using different scales, data on the absolute scale can be log-transformed to be more similar to data on a relative (multiplicative) scale (Wise and Carr 2008b).

## Simulated Vs Natural Herbivory

A majority of studies use simulated or manipulated herbivory, such as clipping leaves or herbivore exclusions, due to the difficulty in controlling damage levels under natural conditions (Tiffin and Inouye 2000). The advantage of using natural herbivory is that plants will experience the pattern of damage that selection has favored tolerance for, but there may be biases resulting from unmeasured environmental variables that may affect both plant and herbivores. Using simulated herbivory allows for the control of environmental variables, but replicating natural herbivory is difficult, causing plants to respond differently from imposed and natural herbivory (Tiffin and Inouye 2000). Growing plants in the control environment of the greenhouse may also affect their response as it is still a novel environment to the plants. Even if the plots are grown in natural settings, the methods of excluding or including herbivores, such as using cages or pesticides, may also affect plant tolerance (Tiffin and Inouye 2000). Lastly, models have predicted that manipulated herbivory may actually result in less precise estimates of tolerance relative to that from natural herbivory (Tiffin and Inoue 2000).

## Fitness Traits

Many studies have shown that using different measurements of fitness may give varying outcomes of tolerance (Strauss and Agrawal 1999; Suwa and Maherali 2008; Banta et al. 2010). Banta et al.

(2010) found that their measure of tolerance will differ depending on whether fruit production or total viable seed production was used to reflect fitness in *Arabdopsis thaliana*. Careful considerations must be made to choose traits that are linked to fitness as closely as possible when measuring tolerance.

## Tolerance-resistance Trade-off

It is classically assumed that there is a negative correlation between the levels of tolerance and resistance in plants (Stowe et al. 2000; Nunez-Farfan et al. 2007). For this trade-off to exist, it requires that tolerance and resistance be redundant defense strategies with similar costs to the plant (Nunez-Farfan et al. 2007). If this is the case, then plants that are able to tolerate damage will suffer little decrease in fitness and so resistance would not be selectively favored. For highly resistant plants, allocating resources to tolerance would not be selectively favored as the plant received minimal damage in the first place.

There is now increasing evidence that many plants allocate resources to both types of defense strategies (Nunez-Farfan et al. 2007). There is also evidence that there may not be a trade-off between tolerance and resistance at all and that they may evolve independently (Leimu and Koricheva 2006; Nunez-Farfan et al. 2007; Muola et al. 2010). Models have shown that intermediate levels of resistance and tolerance are evolutionary stable as long as the benefits of having both traits are more than additive (Nunez-Farfan et al. 2007). Tolerance and resistance may not be redundant strategies since tolerance could be necessary for damage from large mammalian herbivores or specialist herbivores which have the ability to circumvent resistance traits of the plant (Nunez-Farfan et al. 2007; Muola et al. 2010). Also, as traits that confer tolerance are usually basic characteristics of plants, the result of selection on growth and not herbivory may also affect tolerance (Rosenthal and Kotanen 1994).

## Ontogenetic Shifts

It has been suggested that the trade-off between resistance and tolerance may change throughout the development of the plants. It is often assumed that seedlings and juveniles are less tolerant of herbivory since they did not develop the structures required for resource acquisition and so will rely more on traits that confer resistance (Boege et al. 2007; Barton 2008, Barton and Koricheva 2010; Tucker and Avila-Sakar 2010). Although many studies find lower tolerance in seedlings, this is not always the case, as seen in juveniles of *Plant ago lanceolata* which can fully compensate for 50% defoliation (Barton 2008). There is also the added complexity of shifts in herbivore communities as the plant develops and so may favor tolerance or resistance at different life stages (Barton and Koricheva 2010).

## Effects of Resource Levels on Tolerance

The response of plants to herbivory is often plastic and varies according to the conditions it is experiencing (Wise and Abrahamson 2005). The major resources that affect plant growth and also tolerance are water, light, carbon dioxide and soil nutrients. Water and light levels are generally assumed to be positively associated with tolerance (Strauss and Agrawal 1999). However, there are exceptions such as evidence of decreased tolerance in *Madia sativa* with increased water availability (Wise and Abrahamson 2007, Gonzales *et al.* 2008). Many studies have found $CO_2$ levels to decrease tolerance in plants (Lau and Tiffin 2009). Increased nutrient levels are also commonly found to be negatively associated with tolerance (Wise and Abrahamson 2007).

There are currently three prominent models that predict how resource levels may alter a plants 's tolerance to herbivory.

## Growth Rate Model (GRM)

The GRM proposes that the growth rate of the plant at the time of damage is important in determining its response (Hilbert et al. 1981). Plants that are growing in stressful conditions, such as low resource levels or high competition, are growing below their maximum growth rate and so may have a higher capacity for regrowth after receiving damage (Hilbert et al. 1981). In contrast, plants in relatively benign conditions are growing near their maximum growth rate. These plants are less able to recover from damage since they are already near their innate maximum growth rate (Hilbert et al. 1981).

## Compensatory Continuum Hypothesis (CCH)

The CCH suggests that there is a continuum of responses to herbivory (Maschinski and Whitham 1989). It predicts that plants growing in less stressful environment conditions, such as high resource or low competition, are better able to tolerate herbivory since they have abundant resources to replace lost tissues and recover from the damage. Plants growing in stressful environments are then predicted to have lower tolerance (Maschinski and Whitham 1989).

## Limiting Resource Model (LRM)

This recently proposed model takes into account the resource that is limiting plant fitness, the resource affected by herbivory and how the acquisition of resources is affected by herbivory (Wise and Abrahamson 2005). Unlike the GRM and CCH, it is able to incorporate the type of damage received since different modes of herbivory may cause different resources to be affected by herbivory. The LRM encompasses every possible outcome of tolerance (i.e. equal tolerance in both environments, higher tolerance in low stress environments and lower tolerance in low stress environments) and allows multiple pathways to reach these outcome.

Currently, the LRM seems to be most useful in predicting the effects that varying resources levels may have on tolerance (Wise and Abrahamson 2007). Meta-analyses by Hawkes and Sullivan (2001) and Wise and Abrahamson (2007, 2008a) found that the CCH and GRM were insufficient in predicting the diversity of plant tolerance to herbivory. Banta et al. (2010), however, suggested that the LRM should be represented as a set of seven models, instead of one, since each individual part of the LRM requires different assumptions.

## Selection on Herbivores

It is classically assumed that tolerance traits do not impose selection on herbivore fitness (Strauss and Agrawal 1999). This is in contrast to traits that confer resistance, which are likely to affect herbivore fitness and lead to a co-evolutionary arms race (Stinchcombe 2002; Espinosa and Fornoni 2006). However, there are possible mechanisms in which tolerance may affect herbivore fitness .

One mechanism requires a genetic association between loci that confers resistance and tolerance either through tight linkage or pleiotropy (Stinchcombe 2002). Selection for either trait will then also affect the other. If there is a positive correlation between the two traits, then selection for increased tolerance will also increase resistance in the plants. If there is a negative correlation between the two traits then selection for increased tolerance will decrease resistance. How common this association exists, however, is uncertain as there are many studies which find no correlation between tolerance and resistance and others which find significant correlations between them (Leimu and Koricheva 2006; Nunez-Farfan et al. 2007; Muola et al. 2010).

If the traits that allow for tolerance affects the plant tissue's quality, quantity or availability, tolerance may also impose selection on herbivores. Consider a case where tolerance is achieved through activation of dormant meristems in the plants . These new plant tissues may be of lower quality than what was previously eaten by herbivores. herbivores which have higher rates of consumption or can more efficiently use this new resource may be selectively favored over those that cannot (Stinchcombe 2002).

Espinosa and Fornoni (2006) was one study which directly investigated whether tolerance may impose selection on herbivores. As suggested by Stinchcombe (2002), they used plants which had similar resistance but differed in tolerance to more easily differentiate the effects of each trait. As expected, they found evidence that resistance in plants affected herbivore fitness, but they were unable to find any effects of tolerance on herbivore fitness .

A recent model by Restif and Koella (2003) found that plant tolerance can directly impose selection on pathogens. Assuming that investment in tolerance will reduce plant fecundity, infection by pathogens will decrease the number of uninfected hosts. There may then be selection for decreased virulence in the pathogens, so that their plant host will survive long enough to produce enough offspring for future pathogens to infect (Restif and Koelle 2003). However, this may only have limited application to herbivores.

## Species Interactions

### Plant Communities

Herbivory can have large effects on the succession and diversity of plants communities (Anderson and Briske 1995; Stowe et al. 2000; Pejman et al. 2009). Thus, plant defense strategies are important in determining temporal and spatial variation of plant species as it may change the competitive abilities of plants following herbivory. (Anderson and Briske 1995; Stowe et al. 2000).

Past studies have suggested plant resistance to play the major role in species diversity within communities, but tolerance may also be an important factor (Stowe et al. 2000; Pejman et al. 2009). Herbivory may allow less competitive, but tolerant plants to survive in communities dominated by highly competitive but intolerant plant species, thereby increasing diversity (Mariotte et al. 2013). Pejman et al. (2009) found support for this idea in an experimental study on grassland species. In low resource environments, highly competitive (dominant) plants species had lower tolerance than the less competitive (subordinate) species. They also found that the addition of fertilizers offset the negative effects of herbivory on dominant plants. It has also been suggested that the observation of species that occur late in ecological succession (late-seral) being replaced by species that occur in the middle of ecological succession (mid-seral) after high herbivory is due to differences in tolerance between them (Anderson and Briske 1995; Off and Ritchie 1998). However, tolerance between these two groups of species do not always differ and other factors, such as selective herbivory on late-seral species, may contribute to these observations (Anderson and Briske 1995).

## Mutualisms

The large number of studies indicating overcompensation in plants following herbivory, especially after apical meristem damage, has led some authors to suggest that there may be mutualistic relationships between plants and herbivores (Belsky 1986; Agrawal 2000; Edwards 2009). If herbivores provide some benefit for the plant despite causing damage, the plant may evolve tolerance to minimize the damage imposed by the herbivore to shift the relationship more towards mutualism (Edwards 2009). Such benefits include the release from apical dominance, inducing resistance traits to temporally separate herbivores, providing information of future attacks and pollination (Agrawal 2000).

Ipomopsis aggregata

One of the best examples occurs in *Ipomopsis aggregata* where there is increased seed production and seed siring in damaged plants compared to undamaged plants (Figure 4; Edwards 2009). The probability of attack after the first bout of herbivory is low in the environment inhabited by *I. aggregata*. Due to the predictability of attacks, these plants have evolved to overcompensate for the damage and produce the majority of their seeds after the initial bout of herbivory (Edwards 2009). Another example involves endophytic fungi, such as *Neophtodium*, which parasitize plants and produce spores that destroy host inflorescences (Edwards 2009). The fungi also produce alkaloids which protect the plant from herbivores and so the plant may have evolved tolerance to flower damage to acquire this benefit (Edwards 2009). Tolerance may also be involved in the mutualism between the myremecophyte, *Cordia nodosa*, and its ant symbiont *Allomerus octoarticulatus* (Edwards and Yu, 2008). The plant provides the ant with shelter and food bodies in return for protection against herbivory, but the ants also sterilize the plant by removing flower buds. *C. nodosa* is able to compensate for this by reallocating resources to produce flowers on branches not occupied by castrating ants (Edwards and Yu, 2008).

A similar type of mutualism involves plants and mycorrhizal fungi (Bennett and Bever 2007). Mycorrhizal fungi inhabit plant roots and increase nutrient uptake for the plant in exchange for food resources. These fungi are also able to alter the tolerance of plants to herbivory and may cause undercompesation, full compensation and overcompensation depending of the species of fungi involved (Bennett and Bever 2007).

## Agriculture

Modern agriculture has focuses on using genetically modified crops which possess toxic compounds to reduce damage by pests (Nunez-Farfan et al. 2007). However, the effectiveness of resistance traits may decrease as herbivores evolve counter adaptations to the toxic compound, especially since most farmers are reluctant to assign a proportion of their land to contain susceptible crops (Nunez-Farfan et al. 2007). Another method to increase crop yield is to use lines that are tolerant to herbivory and can compensate or even overcompensate for the damage inflicted (Nunez-Farfan et al. 2007; Poveda et al. 2010).

Alterations in resource allocation due to herbivory is studied heavily in agricultural systems (Trumble et al. 1993). Domestication of plants by selecting for higher yield have undoubtedly also caused changes in various plant growth traits, such as decreased resource allocation to non-yield tissues (Welter and Steggall 1993). Alterations in growth traits is likely to effect plant tolerance since the mechanisms overlap. That domesticated tomato plants have lower tolerance to folivory than their wild progenitors suggests this as well (Welter and Steggall 1993). Most agricultural studies however, are more focused on comparing tolerance between damaged and undamaged crops, not between crops and their wild counterparts. Many have found crops, such as cucumbers, cabbages and cauliflowers, can fully compensate and overcompensate for the damaged received (Trumble et al. 1993). A recent study by Poveda et al. (2010) also found evidence of overcompensation in potato plants in response to tuber damage by the potato tuber moth, *Phthorimaea operculella*. Unlike previous examples, the potato plant does not reallocate resources, but actually increases overall productivity to increase mass of tubers and aboveground tissues (Poveda et al. 2010).

## Plant Defense Against Herbivory

Foxgloves produce several deadly chemicals, namely cardiac and steroidal glycosides. Ingestion can cause nausea, vomiting, hallucinations, convulsions, or death.

Plant defense against herbivory or host-plant resistance (HPR) describes a range of adaptations evolved by plants which improve their survival and reproduction by reducing the impact of herbivores. Plants can sense being touched, and they can use several strategies to defend against damage caused by herbivores. Many plants produce secondary metabolites, known as allelochemicals, that influence the behavior, growth, or survival of herbivores. These chemical defenses can act as repellents or toxins to herbivores, or reduce plant digestibility.

Other defensive strategies used by plants include escaping or avoiding herbivores in any time and/or any place, for example by growing in a location where plants are not easily found or accessed by herbivores, or by changing seasonal growth patterns. Another approach diverts herbivores toward eating non-essential parts, or enhances the ability of a plant to recover from the damage caused by herbivory. Some plants encourage the presence of natural enemies of herbivores, which in turn protect the plant. Each type of defense can be either *constitutive* (always present in the plant), or *induced* (produced in reaction to damage or stress caused by herbivores).

Historically, insects have been the most significant herbivores, and the evolution of land plants is closely associated with the evolution of insects. While most plant defenses are directed against insects, other defenses have evolved that are aimed at vertebrate herbivores, such as birds and mammals. The study of plant defenses against herbivory is important, not only from an evolutionary view point, but also in the direct impact that these defenses have on agriculture, including human and livestock food sources; as beneficial 'biological control agents' in biological pest control programs; as well as in the search for plants of medical importance.

## Evolution of Defensive Traits

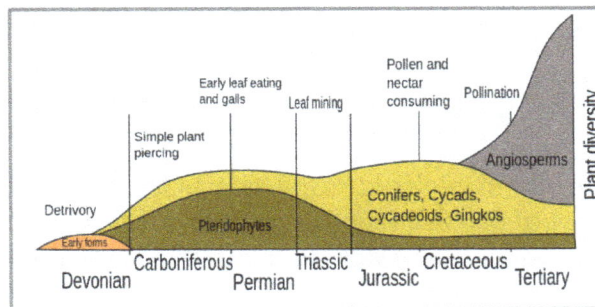

Timeline of plant evolution and the beginnings of different modes of insect herbivory

The earliest land plants evolved from aquatic plants around 450 million years ago (Ma) in the Ordovician period. Many plants have adapted to iodine-deficient terrestrial environment by removing iodine from their metabolism, in fact iodine is essential only for animal cells. An important antiparasitic action is caused by the block of the transport of iodide of animal cells inhibiting sodium-iodide symporter (NIS). Many plant pesticides are glycosides (as the cardiac digitoxin) and cyanogenic glycosides which liberate cyanide, which, blocking cytochrome c oxidase and NIS, is poisonous only for a large part of parasites and herbivores and not for the plant cells in which it seems useful in seed dormancy phase. Iodide is not pesticide, but is oxidized, by vegetable peroxidase, to iodine, which is a strong oxidant, it is able to kill bacteria, fungi and protozoa.

The Cretaceous period saw the appearance of more plant defense mechanisms. The diversification of flowering plants (angiosperms) at that time is associated with the sudden burst of speciation in insects.

This diversification of insects represented a major selective force in plant evolution, and led to selection of plants that had defensive adaptations. Early insect herbivores were mandibulate and bit or chewed vegetation; but the evolution of vascular plants lead to the co-evolution of other forms of herbivory, such as sap-sucking, leaf mining, gall forming and nectar-feeding.

The relative abundance of different species of plants in ecological communities including forests and grasslands may be determined in part by the level of defensive compounds in the different species. Since the cost of replacement of damaged leaves is higher in conditions where resources are scarce, it may also be that plants growing in areas where water and nutrients are scarce may invest more resources into anti-herbivore defenses.

## Records of Herbivores

*Viburnum lesquereuxii* leaf with insect damage; Dakota Sandstone (Cretaceous) of Ellsworth County, Kansas. Scale bar is 10 mm.

Our understanding of herbivory in geological time comes from three sources: fossilized plants, which may preserve evidence of defense (such as spines), or herbivory-related damage; the observation of plant debris in fossilised animal faeces; and the construction of herbivore mouthparts.

Long thought to be a Mesozoic phenomenon, evidence for herbivory is found almost as soon as fossils which could show it. Within under 20 million years of the first fossils of sporangia and stems towards the close of the Silurian, around 420 million years ago, there is evidence that they were being consumed. Animals fed on the spores of early Devonian plants, and the Rhynie chert also provides evidence that organisms fed on plants using a "pierce and suck" technique. Many plants of this time are preserved with spine-like enations, which may have performed a defensive role before being co-opted to develop into leaves.

During the ensuing 75 million years, plants evolved a range of more complex organs - from roots to seeds. There was a gap of 50 to 100 million years between each organ evolving, and it being fed upon. Hole feeding and skeletonization are recorded in the early Permian, with surface fluid feeding evolving by the end of that period.

A Plain Tiger *Danaus chrysippus* caterpillar making a moat to block defensive chemicals of *Calotropis* before feeding

## Co-evolution

Herbivores are dependent on plants for food, and have evolved mechanisms to obtain this food despite the evolution of a diverse arsenal of plant defenses. Herbivore adaptations to plant defense have been likened to *offensive traits* and consist of adaptations that allow increased feeding and use of a host plant. Relationships between herbivores and their host plants often results in recip- rocal evolutionary change, called co-evolution. When an herbivore eats a plant it selects for plants that can mount a defensive response. In cases where this relationship demonstrates *specificity* (the evolution of each trait is due to the other), and *reciprocity* (both traits must evolve), the species are thought to have co-evolved. The "escape and radiation" mechanism for co-evolution presents the idea that adaptations in herbivores and their host plants have been the driving force behind speciation, and have played a role in the radiation of insect species during the age of angiosperms. Some herbivores have evolved ways to hijack plant defenses to their own benefit, by sequestering these chemicals and using them to protect themselves from predators. Plant defenses against her- bivores are generally not complete so plants also tend to evolve some tolerance to herbivory.

## Types

Plant defenses can be classified generally as constitutive or induced. Constitutive defenses are always present in the plant, while induced defenses are produced or mobilized to the site where a plant is injured. There is wide variation in the composition and concentration of constitutive defenses and these range from mechanical defenses to digestibility reducers and toxins. Many ex- ternal mechanical defenses and large quantitative defenses are constitutive, as they require large amounts of resources to produce and are difficult to mobilize. A variety of molecular and biochem- ical approaches are used to determine the mechanism of constitutive and induced plant defenses responses against herbivory.

Induced defenses include secondary metabolic products, as well as morphological and physio- logical changes. An advantage of inducible, as opposed to constitutive defenses, is that they are only produced when needed, and are therefore potentially less costly, especially when herbivory is variable.

## Chemical Defenses

Persimmon, genus *Diospyros*, has a high tannin content which gives immature fruit, seen above, an astringent and bitter flavor.

The evolution of chemical defences in plants is linked to the emergence of chemical substances that are not involved in the essential photosynthetic and metabolic activities. These substances, secondary metabolites, are organic compounds that are not directly involved in the normal growth, development or reproduction of organisms, and often produced as by-products during the synthesis of primary metabolic products. Although these secondary metabolites have been thought to play a major role in defenses against herbivores, a meta-analysis of recent relevant studies has suggested that they have either a more minimal (when compared to other non-secondary metabolites, such as primary chemistry and physiology) or more complex involvement in defense.

Secondary metabolites are often characterized as either *qualitative* or *quantitative*. Qualitative metabolites are defined as toxins that interfere with an herbivore's metabolism, often by blocking specific biochemical reactions. Qualitative chemicals are present in plants in relatively low concentrations (often less than 2% dry weight), and are not dosage dependent. They are usually small, water-soluble molecules, and therefore can be rapidly synthesized, transported and stored with relatively little energy cost to the plant. Qualitative allelochemicals are usually effective against non-adapted specialists and generalist herbivores.

Quantitative chemicals are those that are present in high concentration in plants (5 – 40% dry weight) and are equally effective against all specialists and generalist herbivores. Most quantitative metabolites are digestibility reducers that make plant cell walls indigestible to animals. The effects of quantitative metabolites are dosage dependent and the higher these chemicals' proportion in the herbivore's diet, the less nutrition the herbivore can gain from ingesting plant tissues. Because they are typically large molecules, these defenses are energetically expensive to produce and maintain, and often take longer to synthesize and transport.

The geranium, for example, produces a unique chemical compound in its petals to defend itself from Japanese beetles. Within 30 minutes of ingestion the chemical paralyzes the herbivore. While the chemical usually wears off within a few hours, during this time the beetle is often consumed by its own predators.

- See Toxalbumin

## Types of Chemical Defences

Plants have evolved many secondary metabolites involved in plant defense, which are collectively known as antiherbivory compounds and can be classified into three sub-groups: nitrogen compounds (including *alkaloids, cyanogenic glycosides, glucosinolates* and *benzoxazinoids*), terpenoids, and phenolics.

Alkaloids are derived from various amino acids. Over 3000 known alkaloids exist, examples include nicotine, caffeine, morphine, cocaine, colchicine, ergolines, strychnine, and quinine. Alkaloids have pharmacological effects on humans and other animals. Some alkaloids can inhibit or activate enzymes, or alter carbohydrate and fat storage by inhibiting the formation phosphodiester bonds involved in their breakdown. Certain alkaloids bind to nucleic acids and can inhibit synthesis of proteins and affect DNA repair mechanisms. Alkaloids can also affect cell membrane and cytoskeletal structure causing the cells to weaken, collapse, or leak, and can affect nerve transmission. Although alkaloids act on a diversity of metabolic systems in humans and other animals, they almost uniformly invoke an aversively bitter taste.

Cyanogenic glycosides are stored in inactive forms in plant vacuoles. They become toxic when herbivores eat the plant and break cell membranes allowing the glycosides to come into contact with enzymes in the cytoplasm releasing hydrogen cyanide which blocks cellular respiration. Glucosinolates are activated in much the same way as cyanogenic glucosides, and the products can cause gastroenteritis, salivation, diarrhea, and irritation of the mouth. Benzoxazinoids, secondary defence metabolites, which are characteristic for grasses (Poaceae), are also stored as inactive glucosides in the plant vacuole. Upon tissue disruption they get into contact with β-glucosidases from the chloroplasts, which enzymatically release the toxic aglucones. Whereas some benzoxazinoids are constitutively present, others are only synthesised following herbivore infestation, and thus, considered inducible plant defenses against herbivory.

The terpenoids, sometimes referred to as isoprenoids, are organic chemicals similar to terpenes, derived from five-carbon isoprene units. There are over 10,000 known types of terpenoids. Most are multicyclic structures which differ from one another in both functional groups, and in basic carbon skeletons. Monoterpenoids, continuing 2 isoprene units, are volatile essential oils such as citronella, limonene, menthol, camphor, and pinene. Diterpenoids, 4 isoprene units, are widely distributed in latex and resins, and can be quite toxic. Diterpenes are responsible for making *Rhododendron* leaves poisonous. Plant steroids and sterols are also produced from terpenoid precursors, including vitamin D, glycosides (such as digitalis) and saponins (which lyse red blood cells of herbivores).

Phenolics, sometimes called *phenols*, consist of an aromatic 6-carbon ring bonded to a hydroxy group. Some phenols have antiseptic properties, while others disrupt endocrine activity. Phenolics range from simple tannins to the more complex flavonoids that give plants much of their red, blue, yellow, and white pigments. Complex phenolics called polyphenols are capable of producing many different types of effects on humans, including antioxidant properties. Some examples of phenolics used for defense in plants are: lignin, silymarin and cannabinoids. Condensed tannins, polymers composed of 2 to 50 (or more) flavonoid molecules, inhibit herbivore digestion by binding to consumed plant proteins and making them more difficult for animals to digest, and by interfering with protein absorption and digestive enzymes. Silica and lignins, which are completely indigestible to animals, grind down insect mandibles (appendages necessary for feeding).

In addition to the three larger groups of substances mentioned above, fatty acid derivates, amino acids and even peptides are also used as defense. The cholinergic toxine, cicutoxin of water hemlock, is a polyyne derived from the fatty acid metabolism. β-N-Oxalyl-L-α,β-diaminopropionic acid as simple amino acid is used by the sweet pea which leads also to intoxication in humans. The synthesis of fluoroacetate in several plants is an example of the use of small molecules to disrupt the metabolism of herbivores, in this case the citric acid cycle.

In tropical *Sargassum* and *Turbinaria* species that are often preferentially consumed by herbivorous fishes and echinoids, there is a relatively low level of phenolics and tannins.

## Mechanical Defenses

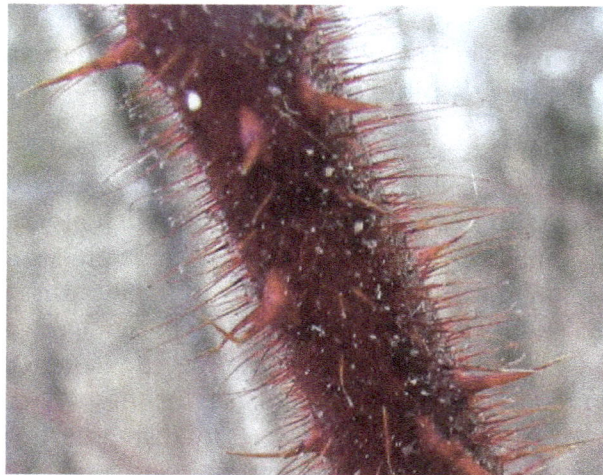

The thorns on the stem of this raspberry plant, serve as a mechanical defense against herbivory.

Plants have many external structural defenses that discourage herbivory. Depending on the herbivore's physical characteristics (i.e. size and defensive armor), plant structural defenses on stems and leaves can deter, injure, or kill the grazer. Some defensive compounds are produced internally but are released onto the plant's surface; for example, resins, lignins, silica, and wax cover the epidermis of terrestrial plants and alter the texture of the plant tissue. The leaves of holly plants, for instance, are very smooth and slippery making feeding difficult. Some plants produce gummosis or sap that traps insects.

A plant's leaves and stem may be covered with sharp prickles, spines, thorns, or trichomes- hairs on the leaf often with barbs, sometimes containing irritants or poisons. Plant structural features like spines and thorns reduce feeding by large ungulate herbivores (e.g. kudu, impala, and goats) by restricting the herbivores' feeding rate, or by wearing down the molars. Raphides are sharp needles of calcium oxalate or calcium carbonate in plant tissues, making ingestion painful, damaging a herbivore's mouth and gullet and causing more efficient delivery of the plant's toxins. The structure of a plant, its branching and leaf arrangement may also be evolved to reduce herbivore impact. The shrubs of New Zealand have evolved special wide branching adaptations believed to be a response to browsing birds such as the moas. Similarly, African Acacias have long spines low in the canopy, but very short spines high in the canopy, which is comparatively safe from herbivores such as giraffes.

Coconut palms protect their fruit by surrounding it with multiple layers of armor.

Trees such as coconut and other palms, may protect their fruit by multiple layers of armor, needing efficient tools to break through to the seed contents, and special skills to climb the tall and relatively smooth trunk.

## Thigmonasty

Thigmonastic movements, those that occur in response to touch, are used as a defense in some plants. The leaves of the sensitive plant, *Mimosa pudica*, close up rapidly in response to direct touch, vibration, or even electrical and thermal stimuli. The proximate cause of this mechanical response is an abrupt change in the turgor pressure in the pulvini at the base of leaves resulting from osmotic phenomena. This is then spread via both electrical and chemical means through the plant; only a single leaflet need be disturbed.

This response lowers the surface area available to herbivores, which are presented with the underside of each leaflet, and results in a wilted appearance. It may also physically dislodge small herbivores, such as insects.

## Mimicry and Camouflage

Some plants mimic the presence of insect eggs on their leaves, dissuading insect species from laying their eggs there. Because female butterflies are less likely to lay their eggs on plants that already have butterfly eggs, some species of neotropical vines of the genus *Passiflora* (Passion flowers) contain physical structures resembling the yellow eggs of *Heliconius* butterflies on their leaves, which discourage oviposition by butterflies.

## Indirect Defenses

The large thorn-like stipules of *Acacia collinsii* are hollow and offer shelter for ants, which in return protect the plant against herbivores.

Another category of plant defenses are those features that indirectly protect the plant by enhancing the probability of attracting the natural enemies of herbivores. Such an arrangement is known as mutualism, in this case of the "enemy of my enemy" variety. One such feature are semiochemicals, given off by plants. Semiochemicals are a group of volatile organic compounds involved in interactions between organisms. One group of semiochemicals are allelochemicals; consisting of allomones, which play a defensive role in interspecies communication, and kairomones, which are used by members of higher trophic levels to locate food sources. When a plant is attacked it releases allelochemics containing an abnormal ratio of these herbivore-induced plant volatiles (HIPVs). Predators sense these volatiles as food cues, attracting them to the damaged plant, and to feeding herbivores. The subsequent reduction in the number of herbivores confers a fitness benefit to the plant and demonstrates the indirect defensive capabilities of semiochemicals. Induced volatiles also have drawbacks, however; some studies have suggested that these volatiles also attract herbivores.

Plants also provide housing and food items for natural enemies of herbivores, known as "biotic" defense mechanisms, as a means to maintain their presence. For example, trees from the genus *Macaranga* have adapted their thin stem walls to create ideal housing for an ant species (genus *Crematogaster*), which, in turn, protects the plant from herbivores. In addition to providing housing, the plant also provides the ant with its exclusive food source; from the food bodies produced by the plant. Similarly, some *Acacia* tree species have developed thorns that are swollen at the base, forming a hollowing structure that acts as housing. These *Acacia* trees also produce nectar in extrafloral nectaries on their leaves as food for the ants.

Plant use of endophytic fungi in defense is a very common phenomenon. Most plants have endophytes, microbial organisms that live within them. While some cause disease, others protect plants from herbivores and pathogenic microbes. Endophytes can help the plant by producing toxins harmful to other organisms that would attack the plant, such as alkaloid producing fungi which are common in grasses such as tall fescue (*Festuca arundinacea*).

## Leaf Shedding and Color

There have been suggestions that leaf shedding may be a response that provides protection against diseases and certain kinds of pests such as leaf miners and gall forming insects. Other responses such as the change of leaf colors prior to fall have also been suggested as adaptations that may help undermine the camouflage of herbivores. Autumn leaf color has also been suggested to act as an honest warning signal of defensive commitment towards insect pests that migrate to the trees in autumn.

## Costs and Benefits

Defensive structures and chemicals are costly as they require resources that could otherwise be used by plants to maximize growth and reproduction. Many models have been proposed to explore how and why some plants make this investment in defenses against herbivores.

## Optimal Defense Hypothesis

The optimal defense hypothesis attempts to explain how the kinds of defenses a particular plant might use reflect the threats each individual plant faces. This model considers three main factors, namely: risk of attack, value of the plant part, and the cost of defense.

The first factor determining optimal defense is risk: how likely is it that a plant or certain plant parts will be attacked? This is also related to the *plant apparency hypothesis*, which states that a plant will invest heavily in broadly effective defenses when the plant is easily found by herbivores. Examples of apparent plants that produce generalized protections include long-living trees, shrubs, and perennial grasses. Unapparent plants, such as short-lived plants of early successional stages, on the other hand, preferentially invest in small amounts of qualitative toxins that are effective against all but the most specialized herbivores.

The second factor is the value of protection: would the plant be less able to survive and reproduce after removal of part of its structure by a herbivore? Not all plant parts are of equal evolutionary value, thus valuable parts contain more defenses. A plant's stage of development at the time of feeding also affects the resulting change in fitness. Experimentally, the fitness value of a plant structure is determined by removing that part of the plant and observing the effect. In general, reproductive parts are not as easily replaced as vegetative parts, terminal leaves have greater value than basal leaves, and the loss of plant parts mid-season has a greater negative effect on fitness than removal at the beginning or end of the season. Seeds in particular tend to be very well protected. For example, the seeds of many edible fruits and nuts contain cyanogenic glycosides such as amygdalin. This results from the need to balance the effort needed to make the fruit attractive to animal dispersers while ensuring that the seeds are not destroyed by the animal.

The final consideration is cost: how much will a particular defensive strategy cost a plant in energy and materials? This is particularly important, as energy spent on defense cannot be used for other functions, such as reproduction and growth. The optimal defense hypothesis predicts that plants will allocate more energy towards defense when the benefits of protection outweigh the costs, specifically in situations where there is high herbivore pressure.

## Carbon:nutrient Balance Hypothesis

The carbon:nutrient balance hypothesis, also known as the *environmental constraint hypothesis* or *Carbon Nutrient Balance Model* (CNBM), states that the various types of plant defenses are responses to variations in the levels of nutrients in the environment. This hypothesis predicts the Carbon/Nitrogen ratio in plants determines which secondary metabolites will be synthesized. For example, plants growing in nitrogen-poor soils will use carbon-based defenses (mostly digestibility reducers), while those growing in low-carbon environments (such as shady conditions) are more likely to produce nitrogen-based toxins. The hypothesis further predicts that plants can change their defenses in response to changes in nutrients. For example, if plants are grown in low-nitrogen conditions, then these plants will implement a defensive strategy composed of constitutive carbon-based defenses. If nutrient levels subsequently increase, by for example the addition of fertilizers, these carbon-based defenses will decrease.

## Growth Rate Hypothesis

The growth rate hypothesis, also known as the *resource availability hypothesis*, states that defense strategies are determined by the inherent growth rate of the plant, which is in turn determined by the resources available to the plant. A major assumption is that available resources are the limiting factor in determining the maximum growth rate of a plant species. This model predicts that the level of defense investment will increase as the potential of growth decreases. Additionally, plants in resource-poor areas, with inherently slow-growth rates, tend to have long-lived leaves and twigs, and the loss of plant appendages may result in a loss of scarce and valuable nutrients.

A recent test of this model involved a reciprocal transplants of seedlings of 20 species of trees between clay soils (nutrient rich) and white sand (nutrient poor) to determine whether trade-offs between growth rate and defenses restrict species to one habitat. When planted in white sand and protected from herbivores, seedlings originating from clay outgrew those originating from the nutrient-poor sand, but in the presence of herbivores the seedlings originating from white sand performed better, likely due to their higher levels of constitutive carbon-based defenses. These finding suggest that defensive strategies limit the habitats of some plants.

## Growth-differentiation Balance Hypothesis

The growth-differentiation balance hypothesis states that plant defenses are a result of a tradeoff between "growth-related processes" and "differentiation-related processes" in different environments. Differentiation-related processes are defined as "processes that enhance the structure or function of existing cells (i.e. maturation and specialization)." A plant will produce chemical defenses only when energy is available from photosynthesis, and plants with the highest concentrations of secondary metabolites are the ones with an intermediate level of available resources. The GDBH also accounts for tradeoffs between growth and defense over a resource availability gradient. In situations where resources (e.g. water and nutrients) limit photosynthesis, carbon supply is predicted to limit both growth and defense. As resource availability increases, the requirements needed to support photosynthesis are met, allowing for accumulation of carbohydrate in tissues. As resources are not sufficient to meet the large demands of growth, these carbon compounds can instead be partitioned into the synthesis of carbon based secondary metabolites (phenolics, tannins, etc.). In environments where the resource demands for growth are met, carbon is allocated to rapidly dividing meristems (high sink strength) at the expense of secondary metabolism. Thus rapidly growing plants are predicted to contain lower levels of secondary metabolites and vice versa. In addition, the tradeoff predicted by the GDBH may change over time, as evidenced by a recent study on Salix spp. Much support for this hypothesis is present in the literature, and some scientists consider the GDBH the most mature of the plant defense hypotheses.

## Importance to Humans

### Agriculture

The variation of plant susceptibility to pests was probably known even in the early stages of agriculture in humans. In historic times, the observation of such variations in susceptibility have provided solutions for major socio-economic problems. The grape phylloxera was introduced from North America to France in 1860 and in 25 years it destroyed nearly a third (100,000 km$^2$) of the French grape yards. Charles Valentine Riley noted that the American species *Vitis labrusca* was resistant to *Phylloxera*. Riley, with J. E. Planchon, helped save the French wine industry by suggesting the grafting of the susceptible but high quality grapes onto *Vitis labrusca* root stocks. The formal study of plant resistance to herbivory was first covered extensively in 1951 by Reginald (R.H.) Painter, who is widely regarded as the founder of this area of research, in his book *Plant Resistance to Insects*. While this work pioneered further research in the US, the work of Chesnokov was the basis of further research in the USSR.

Fresh growth of grass is sometimes high in prussic acid content and can cause poisoning of grazing livestock. The production of cyanogenic chemicals in grasses is primarily a defense against herbivores.

The human innovation of cooking may have been particularly helpful in overcoming many of the defensive chemicals of plants. Many enzyme inhibitors in cereal grains and pulses, such as trypsin inhibitors prevalent in pulse crops, are denatured by cooking, making them digestible.

It has been known since the late 17th century that plants contain noxious chemicals which are avoided by insects. These chemicals have been used by man as early insecticides; in 1690 nicotine was extracted from tobacco and used as a contact insecticide. In 1773, insect infested plants were treated with nicotine fumigation by heating tobacco and blowing the smoke over the plants. The flowers of *Chrysanthemum* species contain pyrethrin which is a potent insecticide. In later years, the applications of plant resistance became an important area of research in agriculture and plant breeding, particularly because they can serve as a safe and low-cost alternative to the use of pesticides. The important role of secondary plant substances in plant defense was described in the late 1950s by Vincent Dethier and G.S. Fraenkel. The use of botanical pesticides is widespread and notable examples include Azadirachtin from the neem (*Azadirachta indica*), d-Limonene from Citrus species, Rotenone from *Derris*, Capsaicin from chili pepper and Pyrethrum.

Natural materials found in the environment also induce plant resistance as well. Chitosan derived from chitin induce a plant's natural defense response against pathogens, diseases and insects including cyst nematodes, both are approved as biopesticides by the EPA to reduce the dependence on toxic pesticides.

The selective breeding of crop plants often involves selection against the plant's intrinsic resistance strategies. This makes crop plant varieties particularly susceptible to pests unlike their wild relatives. In breeding for host-plant resistance, it is often the wild relatives that provide the source of resistance genes. These genes are incorporated using conventional approaches to plant breeding, but have also been augmented by recombinant techniques, which allow introduction of genes from completely unrelated organisms. The most famous transgenic approach is the introduction of genes from the bacterial species, *Bacillus thuringiensis*, into plants. The bacterium produces proteins that, when ingested, kill lepidopteran caterpillars. The gene encoding for these highly toxic proteins, when introduced into the host plant genome, confers resistance against caterpillars, when the same toxic proteins are produced within the plant. This approach is controversial, however, due to the possibility of ecological and toxicological side effects.

## Pharmaceutical

Illustration from the 15th-century manuscript *Tacuinum Sanitatis* detailing the beneficial and harmful properties of Mandrakes

Many currently available pharmaceuticals are derived from the secondary metabolites plants use to protect themselves from herbivores, including opium, aspirin, cocaine, and atropine. These chemicals have evolved to affect the biochemistry of insects in very specific ways. However, many of these biochemical pathways are conserved in vertebrates, including humans, and the chemicals act on human biochemistry in ways similar to that of insects. It has therefore been suggested that the study of plant-insect interactions may help in bioprospecting.

There is evidence that humans began using plant alkaloids in medical preparations as early as 3000 B.C. Although the active components of most medicinal plants have been isolated only recently (beginning in the early 19th century) these substances have been used as drugs throughout the human history in potions, medicines, teas and as poisons. For example, to combat herbivory by the larvae of some Lepidoptera species, Cinchona trees produce a variety of alkaloids, the most familiar of which is quinine. Quinine is extremely bitter, making the bark of the tree quite unpalatable, it is also an anti-fever agent, known as Jesuit's bark, and is especially useful in treating malaria.

Throughout history mandrakes (*Mandragora officinarum*) have been highly sought after for their reputed aphrodisiac properties. However, the roots of the mandrake plant also contain large quantities of the alkaloid scopolamine, which, at high doses, acts as a central nervous system depressant, and makes the plant highly toxic to herbivores. Scopolamine was later found to be medicinally used for pain management prior to and during labor; in smaller doses it is used to prevent motion sickness. One of the most well-known medicinally valuable terpenes is an anticancer drug, taxol, isolated from the bark of the Pacific yew, *Taxus brevifolia*, in the early 1960s.

## Biological Pest Control

Repellent companion planting, defensive live fencing hedges, and "obstructive-repellent" interplanting, with host-plant resistance species as beneficial 'biological control agents' is a technique in biological pest control programs for: organic gardening, wildlife gardening, sustainable gardening, and sustainable landscaping; in organic farming and sustainable agriculture; and in restoration ecology methods for habitat restoration projects.

# Inducible Plant Defenses Against Herbivory

Plants and herbivores have co-evolved together for 350 million years. Plants have evolved many defense mechanisms against insect herbivory. Such defenses can be broadly classified into two categories: (1) permanent, constitutive defenses, and (2) temporary, inducible defenses. Both types are achieved through similar means but differ in that constitutive defenses are present before an herbivore attacks, while induced defenses are activated only when attacks occur. In addition to constitutive defenses, initiation of specific defense responses to herbivory is an important strategy for plant persistence and survival.

## Benefits of Induced Defences

Inducible defenses allow plants to be phenotypically plastic. This may confer an advantage over constitutive defenses for multiple reasons. First, it may reduce the chance that attacking insects adapt to plant defenses. Simply, inducible defenses cause variations in the defense constituents of a plant, thereby making the plant a more unpredictable environment for insect herbivores. This variability has an important effect on the fitness and behaviour of herbivores. For example, the study of herbivory on radish (*Raphanus sativus*) by the cabbage looper caterpillar (*Trichoplusia ni*) demonstrated that the variation of defensive chemicals (glucosinolates) in *R. sativus*, due to induction, resulted in a significant decrease in the pupation rates of *T. ni*. In essence, defensive chemicals can be viewed as having a particular dosage-dependent effect on herbivores: it has little detrimental effect on herbivores when present at a low or moderate dose, but has dramatic effects at higher concentrations. Hence, a plant which produces variable levels of defensive chemicals is better defended than one that always produces the mean level of toxin.

Second, synthesizing a continually high level of defensive chemicals renders a cost to the plant. This is particularly the case where the presence of herbivorous insects is not always predictable. For example, the production of nicotine in cultivated tobacco (*Nicotiana tabacum*) has a function in plant defence. *N. tabacum* plants with a higher constitutive level of nicotine are less susceptible to insect herbivory. However, *N. tabacum* plants that produce a continually high level of nicotine flower significantly later than plants with lower levels of nicotine. This results suggest that there is a biosynthetic cost to constantly producing a high level of defensive chemicals. Inducible defences are advantageous as they reduce the metabolic load on the plant in conditions where such biological chemicals are not yet necessary. This is particularly the case for defensive chemicals containing nitrogen (e.g. alkaloids) as if the plant is not being attacked it is able to divert more nitrogen to producing rubisco and will therefore be able to grow faster and produce more seeds.

In addition to chemical defenses, herbivory can induced physical defenses, such as longer thorns, or indirect defenses, such as rewards for symbiotic ants.

## Cost of Induced Defences

Central to the concept of induced defences is the cost involved when stimulating such defences in the absence of insect herbivores. After all, in the absence of cost, selection is expected to favour the most defended genotype. Accordingly, individual plants will only do so when there is a need to. The cost of induced defences to a plant can be quantified as the resource-based trade-off between resistance and fitness (allocation cost) or as the reduced fitness resulting from the interactions with other species or the environment (ecological cost).

Allocation cost is related to the channelling of a large quantity fitness-limited resources to from resistance traits in plants. Such resources might not be quickly recycled and thus, are unavailable for fitness-relevant process such as growth and reproduction. For instance, herbivory on the broadleaf dock (*Rumex obtusifolius*) by the green dock beetle (*Gastrophysa viridula*) induces an increased activity in cell wall-bound peroxidase. The allocation of resources to this increased activity results in reduced leaf growth and expansion in *R. obtusifolius*. In the absence of herbivory, inducing such a defence would be ultimately costly to the plant in terms of development.

Ecological cost results from the disruption of the many symbiotic relationships that a plant has with the environment. For example, jasmonic acid can be used to simulate an herbivore attack on plants and thus, induce plant defences. The use of jasmonic acid on tomato (*Lycopersicon esculentum*) resulted in plants with fewer but larger fruits, longer ripening time, delayed fruit-set, fewer seeds per plant and fewer seeds per unit of fruit weight. All these features play a critical role in attracting seed dispersers. Due to the consequences of induced defences on fruit characteristics, *L. esculentum* are less able to attract seed dispersers and this ultimately results in a reduced fitness.

## Sensing Herbivory Attack

Induced defences require plant sensing the nature of injury, such as wounding from herbivore attack as opposed to wounding from mechanical damage. Plants therefore use a variety of cues, including the sense of touch, and salivary enzymes of the attacking herbivore. For example, in a study to test whether plants can distinguish mechanical damage from insect herbivory attack, Korth and Dixon (1997) discovered that the accumulation of induce defence transcription products occurred more rapidly in potato (*Solanum tuberosum L.*) leaves chewed on by caterpillars than in leaves damaged mechanically. Distinct signal transduction pathway are activated in response either to insect damage or mechanical damage in plants. While chemicals released in wounding responses are the same in both cases, the pathway in which they accumulate are separate. Not all herbivore attack begins with feeding, but with insects laying eggs on the plant. The adults of butterflies and moths (order Lepidoptera), for example, do not feed on plants directly, but lay eggs on plants which are suitable food for their larva. In such cases, plants have been demonstrated to induce defences upon contact from the ovipositing of insects.

## A mechanism of Defence Induction: Changes in Gene Transcription Rates

Systemically induced defences are at least in some cases the result of changes in the transcription rates of genes in a plant. Genes involved in this process may differ between species, but common to all plants is that systemically induced defences occur as a result of changes in gene expression. The changes in transcription can involve genes which either do not encode products involved in insect resistance, or are involved in general response to stress. In cultivated tobacco (*Nicotiana tobacum*) photosynthetic genes are down-regulated, while genes directly involved in defences are up-regulated in response to insect attack. This allows more resources to be allocated to producing proteins directly involved in the resistance response. A similar response was reported in *Arabidopsis* plants where there is an up-regulation of all genes that are involved in defence. Such changes in the transcription rates are essential in inducing a change in the level of defence upon herbivory attack.

## Classification of Induced Genes

Not all up-regulated genes in induced defences are directly involved in the production of toxins. The genes encoding newly synthesised proteins after a herbivory attack can be categorised based on the function of their transcriptional products. There are three broad classification categories: defence genes, signalling pathway genes and rerouting genes. The transcription of defensive gene produces either proteins that are directly involved in plant defence such as proteinase inhibitors or are enzymes that are essential for the production of such proteins. Signalling pathway genes are involved in transmitting the stimulus from the wounded regions to organs where defence genes are transcribed. These genes are essential in plants due to the constraints in the vascular systems of the plants. Finally, rerouting gene are responsible in allocating resources for metabolism from primary metabolites involved in photosynthesis and survival to defence genes.

# Plant Use of Endophytic Fungi in Defense

Plant use of endophytic fungi in defense occurs when endophytic fungi, which live symbiotically with the majority of plants by entering their cells, are utilized as an indirect defense against herbivores. In exchange for carbohydrate energy resources, the fungus provides benefits to the plant which can include increased water or nutrient uptake and protection from phytophagous insects, birds or mammals. Once associated, the fungi alter nutrient content of the plant and enhance or begin production of secondary metabolites. The change in chemical composition acts to deter herbivory by insects, grazing by ungulates and/or oviposition by adult insects. Endophyte-mediated defense can also be effective against pathogens and non-herbivory damage.

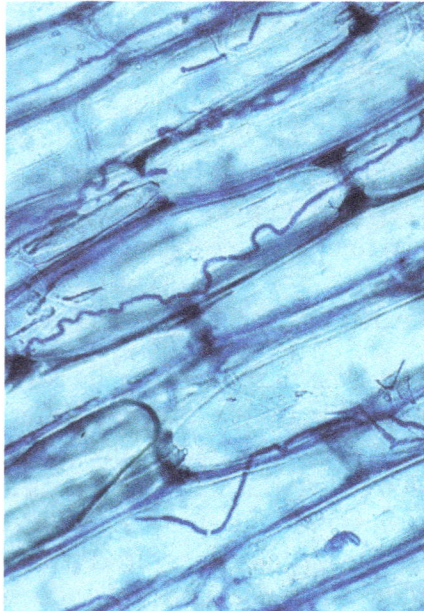

Neotyphodium spp.are commonly associated with tall fescue in the leaf sheath tissue.
They produce secondary metabolites toxic to herbivores.

This differs from other forms of indirect defense in that the fungi live within the plant cells and directly alter their physiology. In contrast, other biotic defenses such as predators or parasites of the herbivores consuming a plant are normally attracted by volatile organic compounds (known as semiochemicals) released following damage or by food rewards and shelter produced by the plant. These defenders vary in the time spent with the plant: from long enough to oviposit to remaining there for numerous generations, as in the ant-acacia mutualism. Endophytic fungi tend to live with the plant over its entire life.

## Diversity of Endophytic Associations

The fungal endophytes are a diverse group of organisms forming associations almost ubiquitously throughout the plant kingdom. The endophytes which provide indirect defense against herbivores may have come from a number of origins, including mutualistic root endophyte associations and the evolution of entomopathogenic fungi into plant-associated endophytes. The endomycorrhiza, which live in plant roots, are made up of five groups: arbuscular, arbutoid, ericoid, monotropoid, and orchid mycorrhizae. The majority of species are from the phylum Glomeromycota with the ericoid species coming from the Ascomycota, while the arbutoid, monotropoid and orchid mycorrhizae are classified as Basidiomycota. The entomopathogenic view has gained support from observations of increased fungal growth in response to induced plant defenses and colonization of plant tissues.

Claviceps spp. fungus growing on wheat spikes, a common endophyte of the grasses.

Examples of host specialists are numerous – especially in temperate environments – with multiple specialist fungi frequently infecting one plant individual simultaneously. These specialists demonstrate high levels of specificity for their host species and may form physiologically adapted host-races on closely related congeners. *Piriformospora indica* is an interesting endophytic fungus of the order Sebacinales, the fungus is capable of colonising roots and forming symbiotic relationship with every possible plant on earth . *P. indica* has also been shown to increase both crop yield and plant defence of a variety of crops(barley, tomato, maize etc.) against root-pathogens. However, there are also many examples of generalist fungi which may occur on different hosts at different frequencies (e.g. Acremonium endophytes from five subgenera of Festuca) and as part of a variety of fungal assemblages. They may even spread to novel, introduced plant species. Endophytic mutualists associate with species representative of every growth form and life history strategy in the grasses and many other groups of plants. The effects of associating with multiple strains or species of fungus at once can vary, but in general, one type of fungus will be providing the majority of benefit to the plant.

## Mechanisms of Defense

## Secondary Metabolite Production

Some chemical defenses once thought to be produced by the plant have since been shown to be synthesized by endophytic fungi. The chemical basis of insect resistance in endophyte-plant defense mutualisms has been most extensively studied in the perennial ryegrass and three major classes of secondary metabolites are found: indole diterpenes, ergot alkaloids and peramine. Related compounds are found across the range of endophytic fungal associations with plants. The terpenes and alkaloids are inducible defenses which act similarly to defensive compounds produced by plants and are highly toxic to a wide variety of phytophagous insects as well as mammalian herbivores. Peramine occurs widely in endophyte-associated grasses and may also act as a signal to invertebrate herbivores of the presence of more dangerous defensive chemicals. Terpenoids and ketones have been linked to protection from specialist and generalist herbivores (both insect and vertebrate) across the higher plants.

Generalist herbivores are more likely than specialists to be negatively affected by the defense chemicals that endophytes produce because they have, on average, less resistance to these specific, qualitative defenses. Among the chewing insects, infection by mycorrhizae can actually benefit specialist feeders even if it negatively affects generalists. The overall pattern of effects on insect herbivores seems to support this, with generalist mesophyll feeders experiencing negative effects of host infection, although phloem feeders appear to be affected little by fungal defenses.

Secondary metabolites may also affect the behaviour of natural enemies of herbivorous species in a multi-trophic defense/predation association. For instance, terpenoid production attracts natural enemies of herbivores to damaged plants. These enemies can reduce numbers of invertebrate herbivores substantially and may not be attracted in the absence of endophytic symbionts. Multi-trophic interactions can have cascading consequences for the entire plant community, with the potential to vary widely depending on the combination of fungal species infecting a given plant and the abiotic conditions.

## Altered Nutrient Content

Due to the inherently nutrient-exchange based economy of the plant-endophyte association, it is not surprising that infection by fungi directly alters the chemical composition of plants, with corresponding impacts on their herbivores. Endophytes frequently increase apoplastic carbohydrate concentration, altering the C:N ratio of leaves and making them a less efficient source of protein. This effect can be compounded when the fungus also uses plant nitrogen to form N-based secondary metabolites such as alkaloids. For example, the thistle gall fly (*Urophora cardui*) experiences reduced performance on plants infected with endophytic fungi due to the decrease in N-content and ability to produce large quantities of high-quality gall tissue. Additionally, increased availability of limiting nutrients to plants improves overall performance and health, potentially increasing the ability of infected plants to defend themselves.

## Impacts on Host Plants

### Herbivory Prevention

Studies of fungal infection consistently reveal that plants with endophytes are less likely to suffer substantial damage, and herbivores feeding on infected plants are less productive. There are multiple modes through which endophytic fungi reduce insect herbivore damage, including avoidance (deterrence), reduced feeding, reduced development rate, reduced growth and/or population growth, reduced survival and reduced oviposition. Vertebrate herbivores such as birds, rabbits and deer show the same patterns of avoidance and reduced performance. Even below-ground herbivores such as nematodes and root-feeding insects are reduced by endophyte infection. The strongest evidence for anti-herbivore benefits of fungal endophytes come from studies of herbivore populations being extirpated when allowed to feed only on infected plants. Examples of local extinction have been documented in crickets, larval armyworms and flour beetles.

Yet chemical defenses produced by fungal endophytes are not universally effective, and numerous insect herbivores are unaffected by a given compound at one or more life history stages; larval stages are often more susceptible to toxins than adults. Even endophytes which purportedly provide some defense benefit to their hosts such as the Neotyphidium partner of many grass species in the alpine tundra do not always lead to avoidance or ill-effects on herbivores due to spatial variation in levels of consumption.

## Mutualism-pathogenicity Continuum

Not all endophytic symbioses confer protection from herbivores – only some species associations act as defense mutualisms. The difference between a mutualistic endophyte and a pathogenic one can be indistinct and dependent on interactions with other species or environmental conditions. Some fungi which are pathogens in the absence of herbivores may become beneficial under high levels of insect damage, such as species which kill plant cells in order to make nutrients available for their own growth, thereby altering nutritional content of leaves and making them a less desirable foodstuff. Some endomycorrhizae may provide defense benefits but at the cost of lost reproductive potential by rendering grasses partially sterile with their own fungal reproductive structures taking precedence. This is not unusual among fungi, as non-endophytic plant pathogens have similar conditionally beneficial effects on defense. Some species of endophyte may be beneficial for the plants in other ways (e.g. nutrient and water uptake) but will provide less benefit as a plant receives more damage and not produce defensive chemicals in response. The effect of one fungus on the plant can be altered when multiple strains of fungi are infecting a given individual in combination.

Some endomycorrhizae may actually promote herbivore damage by making plants more susceptible to it. For example, some oak fungal endophytes are positively correlated with the levels of damage from leaf miners (*Cameraria spp.*), although negatively correlated with number of larvae present due to a reduction of oviposition on infected plants, which partially mitigates the higher damage rate. This continuum between mutualism and pathogenicity of endophytic fungi has major implications for plant fitness depending on the species of partners available in a given environment; mutualist status is conditional in a way similar to pollination and can shift from one to the other just as frequently.

## Fitness and Competitive Ability

Fungal endophytes which provide defensive services to their host plants may exert selective pressures favouring association through enhanced fitness relative to uninfected hosts. The fungus Neotyphodium spp. infects grasses and increases fitness under conditions with high levels of interspecific competition. It does this through a combination of benefits including anti-herbivore defenses and growth promoting factors. The customary assumption that plant growth promotion is the main way fungal mutualists improve fitness under attack from herbivores is changing; alteration of plant chemical composition and induced resistance are now recognized as factors of great importance in improving competitive ability and fecundity. Plants undefended by chemical or physical means at certain points in their life histories have higher survival rates when infected with beneficial endophytic fungi. The general trend of plants infected with mutualistic fungi outperforming uninfected plants under moderate to high herbivory exerts selection for higher levels of fungal association as herbivory levels increase. Unsurprisingly, low to moderate levels of herbivore damage also increases the levels of infection by beneficial endophytic fungi.

In some cases the symbiosis between fungus and plant reaches a point of inseparability; fungal material is transmitted vertically from the maternal parent plant to seeds, forming a near-obligate mutualism. Because seeds are an important aspect of both fecundity and competitive ability for plants, high germination rates and seedling survival increase lifetime fitness. When fitness of plant and fungus become tightly intertwined, it is in the best interest of the endophyte to act in a manner beneficial to the plant, pushing it further toward the mutualism end of the continuum. Such effects of seed defense can also occur in dense stands of conspecifics through horizontal transmission of beneficial fungi. Mechanisms of microbial association defense, protecting the seeds rather than the already established plants, can have such drastic impacts on seed survival that they have been recognized to be an important aspect of the larger 'seed defence theory'.

## Climate Change

The range of associated plants and fungi may be altered as climate changes, and not necessarily in a synchronous fashion. Plants may lose or gain endophytes, with as yet unknown impacts on defense and fitness, although generalist species may provide indirect defense in new habitats more often than not. Above-ground and below-ground associations can be mutual drivers of diversity, so altering the interactions between plants and their fungi may also have drastic effects on the community at large, including herbivores. Changes in distribution may bring plants into competition with previously established local species, making the fungal community – and particularly the pathogenic role of fungus – important in determining outcomes of competition with non-native invasive species. As carbon dioxide levels rise, the amplified photosynthesis will increase the pool of carbohydrates available to endophytic partners, potentially altering the strength of associations. Infected C3 plants show greater relative growth rate under high $CO_2$ conditions compared to uninfected plants, and it is possible that the fungi drive this pattern of increased carbohydrate production.

Levels of herbivory may also increase as temperature and carbon dioxide concentrations rise. However, should plants remain associated with their current symbiotic fungi, evidence suggests that the degree of defense afforded them should not be altered. Although the amount of damage caused by herbivores frequently increases under elevated levels of atmospheric $CO_2$, the proportion of damage remains constant when host plants are infected by their fungal endophytes. The change in Carbon-Nitrogen ratio will also have important consequences for herbivores. As carbohydrate levels increase within plants, relative nitrogen content will fall, having the dual effects of reducing nutritional benefit per unit biomass and also lowering concentrations of nitrogen-based defenses such as alkaloids.

## History of Research

### Early Recognition

The effects of endophytic fungi on the chemical composition of plants have been known by humans for centuries in the form of poisoning and disease as well as medicinal uses. Especially noted were impacts on agricultural products and livestock. Recognition and study of the mutualism did not begin in earnest until the 1980s when early studies on the impacts of alkaloids on animal herbivory confirmed their importance as agents of deterrence. Biologists began to characterize the diversity of endophytic mutualists through primitive techniques such as isozyme analysis and measuring the effects of infection on herbivores. Basic descriptive accounts of these previously neglected species of fungus became a major goal for mycologists, and a lot of research focus shifted to associates of the grass family (Poaceae) in particular, because of the large number of species which represent economically important commodities to humans.

## Recent Advances and Future Directions

In addition to continuing descriptive studies of the effects of infection by defense mutualist endophytes, there has been a sharp increase in the number of studies which delve further into the ecology of plant-fungus associations and especially their multi-trophic impacts. The processes by which endophytic fungi alter plant physiology and volatile chemical levels are virtually unknown, and limited current results show a lack of consistency under differing environmental conditions, especially differing levels of herbivory. Studies comparing the relative impacts of mutualistic endophytes on inducible defenses and tolerance show a central function of infection in determining both responses to herbivore damage.[100] On the whole, molecular mechanisms behind endophyte-mediated plant defense has been an increasing focus of research over the past ten years.

Since the beginning of the biotechnology revolution, much research has been also focused on using genetically modified endophytes to improve plant yields and defensive properties. The genetic basis of response to herbivory is being explored in tall fescue, where it appears the production of jasmonic acid may play a role in downregulation of the host plant's chemical defense pathways when a fungal endophyte is present. In some cases, fungi that are closely associated with their hosts have transferred genes for secondary metabolite production to the host genome, which could help to explain multiple origins of chemical defenses within the phylogeny of various groups of plants. This represents an important line of inquiry to pursue, especially in regards to understanding the chemical pathways that can be utilized in biotechnological applications.

## Importance to Humans

### Agriculture and Livestock

The secondary chemicals produced by endophytic fungi when associated with their host plants can be very harmful to mammals including livestock and humans, causing more than 600 million dollars in losses due to dead livestock every year. For example, the ergot alkaloids produced by Claviceps spp. have been dangerous contaminants of rye crops for centuries. When not lethal, defense chemicals produced by fungal endophytes may lead to lower productivity in cows and other livestock feeding on infected forage. Reduced nutritional quality of infected plant tissue also lowers the performance of farm animals, compounding the effect of reduced feed uptake when provided with infected plant matter. Reduced frequency of pregnancy and birth has also been reported in cattle and horses fed with infected forage. Consequently, the dairy and meat-production industries must endure substantial economic losses.

Fungal resistance to herbivores represents an environmentally sustainable alternative to pesticides that has experienced reasonable success in agricultural applications. The organic farming industry has embraced mycorrhizal symbionts as one tool for improving yields and protecting plants from damage. Infected crops of soybean, ribwort plantain, cabbage, banana, coffee bean plant and tomato all show markedly lower rates of herbivore damage compared to uninfected plants. Endophytic fungi show great promise as a means of indirect biocontrol in large-scale agricultural applications. The potential for biotechnology to improve crop populations through inoculation with modified fungal strains could reduce toxicity to livestock and improve yields of human-consumed foods. The endophyte, either with detrimental genes removed or beneficial new genes added, is used as a surrogate host to transform the crops genetically. An endophyte of ryegrass has been genetically transformed in this way and used successfully to deter herbivores.

Understanding how to mediate top-down effects on crop populations caused by the enemies of herbivores as well as bottom-up effects of chemical composition in infected plants has important consequences for the management of agricultural industries. The selection of endophytes for agricultural use must be careful and consideration must be paid to the specific impacts of infection on all species of pest and predators or parasites, which may vary on a geographic scale. The union of ecological and molecular techniques to increase yield without sacrificing the health of the local or global environment is a growing area of research.

## Pharmaceutical

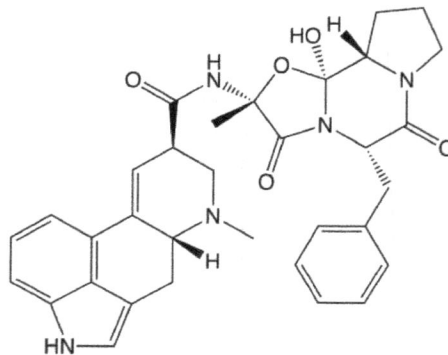

Ergotamine, a mycotoxin produced by Claviceps spp. which infects rye and related grasses, causing poisoning of livestock and humans. It is used in a number of medicinal

Many secondary metabolites from endophyte-plant interactions have also been isolated and used in raw or derived forms to produce a variety of drugs treating many conditions. The toxic properties of ergot alkaloids also make them useful in the treatment of headaches and throughout the process of giving birth by inducing contractions and stemming hemorrhages. Drugs used to treat Parkinson's Disease have been created from isolates of ergot toxins, although health risks may accompany their use. Ergotamine has also been used to synthesize lysergic acid diethylamide because of its chemical similarity to lysergic acid. The generally chemically-based defense properties of endophytic fungi make them a perfect group of organisms to search for new antibiotic compounds within, as other fungi have in the past yielded such useful drugs as penicillin and streptomycin and plants use their antibiotic qualities as a defense against pathogens.

## References

- Norman M. Wereley; Janet M. Sater (2012). Plants and Mechanical Motion: A Synthetic Approach to Nastic Materials and Structures. DEStech Publications, Inc. ISBN 978-1-60595-043-3.

- Dov Koller; Elizabeth Van Volkenburgh (15 January 2011). THE RESTLESS PLANT. Harvard University Press. pp. 18–. ISBN 978-0-674-05943-6.

- Futuyma, Douglas J.; Montgomery Slatkin (1983). Coevolution. Sunderland, Massachusetts: Sinauer Associates. ISBN 0-87893-228-3.

- Whittaker, Robert H. (1970). "The biochemical ecology of higher plants". In Ernest Sondheimer; John B. Simeone. Chemical ecology. Boston: Academic Press. pp. 43–70. ISBN 0-12-654750-5.

- Roberts, Margaret F.; Michael Wink (1998). Alkaloids: biochemistry, ecology, and medicinal applications. New York: Plenum Press. ISBN 0-306-45465-3.

- Gershenzon, Jonathan; Wolfgang Kreis (1999). "Biochemistry of terpenoids". In Michael Wink. Biochemistry of plant secondary metabolism. London: Sheffield Academic Press. pp. 222–279. ISBN 0-8493-4085-3.

- Michael Smith, C. (2005). Plant Resistance to Arthropods: Molecular and Conventional Approaches. Berlin: Springer. ISBN 1-4020-3701-5.

- Peterson, R.L.; Massicotte, H.B. & Melville, L.H. (2004). Mycorrhizas: anatomy and cell biology. National Research Council Research Press. ISBN 978-0-660-19087-7.

- Clay, K. (1994). The potential role of endophytes in ecosystems. In: Biotechnology of endophytic fungi of grasses. Boca Raton: CRC Press. pp. 73–86. ISBN 978-0-8493-6276-7

- Black, M.H. & Halmer, P. (2006). The encyclopedia of seeds: science, technology and uses. Wallingford, UK: CABI. pp. 226. ISBN 978-0-85199-723-0.

- Chapin, F. Stuart, III (1980). "The Mineral Nutrition of Wild Plants". Annual Review of Ecology and Systematics. 11: 233–260. doi:10.1146/annurev.es.11.110180.001313. JSTOR 2096908. Retrieved 2014-01-15.

# Multidisciplinary Fields of Plant Pathology

The chemicals that regulate plant growth are known as plant hormones. Plant hormones occur within the plant and occur in extremely low concentrations. Hormones also determine the formation of flowers, stems, leaves and the shedding of leaves. The following content also focuses on forest pathology, which is the research of both biotic and abiotic maladies affecting the health of a forest ecosystem. Forest pathology is a part of the comprehensive approach of forest protection.

## Plant Hormone

Plant hormones (also known as phytohormones) are chemicals that regulate plant growth. In the United Kingdom, these are termed 'plant growth substances'.

Lack of the plant hormone auxin can cause abnormal growth (right)

Plant hormones are signal molecules produced within the plant, and occur in extremely low concentrations. Hormones regulate cellular processes in targeted cells locally and, moved to other locations, in other functional parts of the plant. Hormones also determine the formation of flowers, stems, leaves, the shedding of leaves, and the development and ripening of fruit. Plants, unlike animals, lack glands that produce and secrete hormones. Instead, each cell is capable of producing hormones. Plant hormones shape the plant, affecting seed growth, time of flowering, the sex of flowers, senescence of leaves, and fruits. They affect which tissues grow upward and which grow downward, leaf formation and stem growth, fruit development and ripening, plant longevity, and even plant death. Hormones are vital to plant growth, and, lacking them, plants would be mostly a mass of undifferentiated cells. So they are also known as growth factors or growth hormones. The term 'Phytohormone' was coined by Thimann in 1948.

Phytohormones are found not only in higher plants, but in algae too, showing similar functions, and in microorganisms, like fungi and bacteria, but, in this case, they play no hormonal or other immediate physiological role in the producing organism and can, thus, be regarded as secondary metabolites.

## Characteristics

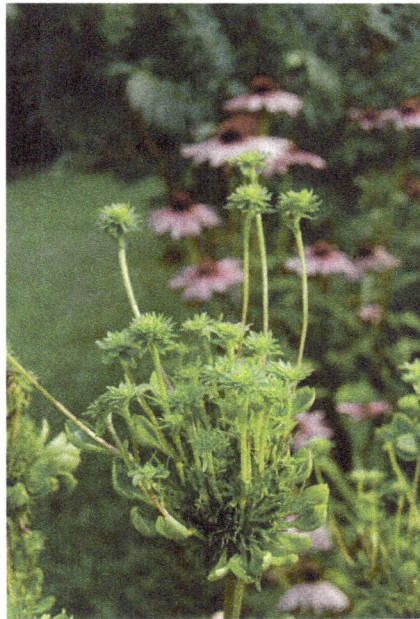

Phyllody on a purple coneflower (*Echinacea purpurea*), a plant development abnormality where leaf-like structures replace flower organs. It can be caused by hormonal imbalance, among other reasons.

The word hormone is derived from Greek, meaning *set in motion*. Plant hormones affect gene expression and transcription levels, cellular division, and growth. They are naturally produced within plants, though very similar chemicals are produced by fungi and bacteria that can also affect plant growth. A large number of related chemical compounds are synthesized by humans. They are used to regulate the growth of cultivated plants, weeds, and in vitro-grown plants and plant cells; these man-made compounds are called Plant Growth Regulators or PGRs for short. Early in the study of plant hormones, "phytohormone" was the commonly used term, but its use is less widely applied now.

Plant hormones are not nutrients, but chemicals that in small amounts promote and influence the growth, development, and differentiation of cells and tissues. The biosynthesis of plant hormones within plant tissues is often diffuse and not always localized. Plants lack glands to produce and store hormones, because, unlike animals — which have two circulatory systems (lymphatic and cardiovascular) powered by a heart that moves fluids around the body — plants use more passive means to move chemicals around the plant. Plants utilize simple chemicals as hormones, which move more easily through the plant's tissues. They are often produced and used on a local basis within the plant body. Plant cells produce hormones that affect even different regions of the cell producing the hormone.

Hormones are transported within the plant by utilizing four types of movements. For localized movement, cytoplasmic streaming within cells and slow diffusion of ions and molecules between cells are utilized. Vascular tissues are used to move hormones from one part of the plant to another; these include sieve tubes or phloem that move sugars from the leaves to the roots and flowers, and xylem that moves water and mineral solutes from the roots to the foliage.

Not all plant cells respond to hormones, but those cells that do are programmed to respond at specific points in their growth cycle. The greatest effects occur at specific stages during the cell's life, with diminished effects occurring before or after this period. Plants need hormones at very specific times during plant growth and at specific locations. They also need to disengage the effects that hormones have when they are no longer needed. The production of hormones occurs very often at sites of active growth within the meristems, before cells have fully differentiated. After production, they are sometimes moved to other parts of the plant, where they cause an immediate effect; or they can be stored in cells to be released later. Plants use different pathways to regulate internal hormone quantities and moderate their effects; they can regulate the amount of chemicals used to biosynthesize hormones. They can store them in cells, inactivate them, or cannibalise already-formed hormones by conjugating them with carbohydrates, amino acids, or peptides. Plants can also break down hormones chemically, effectively destroying them. Plant hormones frequently regulate the concentrations of other plant hormones. Plants also move hormones around the plant diluting their concentrations.

The concentration of hormones required for plant responses are very low ($10^{-6}$ to $10^{-5}$ mol/L). Because of these low concentrations, it has been very difficult to study plant hormones, and only since the late 1970s have scientists been able to start piecing together their effects and relationships to plant physiology. Much of the early work on plant hormones involved studying plants that were genetically deficient in one or involved the use of tissue-cultured plants grown *in vitro* that were subjected to differing ratios of hormones, and the resultant growth compared. The earliest scientific observation and study dates to the 1880s; the determination and observation of plant hormones and their identification was spread-out over the next 70 years.

## Classes of Plant Hormones

In general, it is accepted that there are five major classes of plant hormones, some of which are made up of many different chemicals that can vary in structure from one plant to the next. The chemicals are each grouped together into one of these classes based on their structural similarities and on their effects on plant physiology. Other plant hormones and growth regulators are not easily grouped into these classes; they exist naturally or are synthesized by humans or other organisms, including chemicals that inhibit plant growth or interrupt the physiological processes within plants. Each class has positive as well as inhibitory functions, and most often work in tandem with each other, with varying ratios of one or more interplaying to affect growth regulation.

The five major classes are:

## Abscisic Acid

Abscisic acid (also called ABA) is one of the most important plant growth regulators. It was discovered and researched under two different names before its chemical properties were fully known, it was called *dormin* and *abscicin II*. Once it was determined that the two compounds are the same, it was named abscisic acid. The name "abscisic acid" was given because it was found in high concentrations in newly abscissed or freshly fallen leaves.

This class of PGR is composed of one chemical compound normally produced in the leaves of plants, originating from chloroplasts, especially when plants are under stress. In general, it acts as an inhibitory chemical compound that affects bud growth, and seed and bud dormancy. It mediates changes within the apical meristem, causing bud dormancy and the alteration of the last set of leaves into protective bud covers. Since it was found in freshly abscissed leaves, it was thought to play a role in the processes of natural leaf drop, but further research has disproven this. In plant species from temperate parts of the world, it plays a role in leaf and seed dormancy by inhibiting growth, but, as it is dissipated from seeds or buds, growth begins. In other plants, as ABA levels decrease, growth then commences as gibberellin levels increase. Without ABA, buds and seeds would start to grow during warm periods in winter and be killed when it froze again. Since ABA dissipates slowly from the tissues and its effects take time to be offset by other plant hormones, there is a delay in physiological pathways that provide some protection from premature growth. It accumulates within seeds during fruit maturation, preventing seed germination within the fruit, or seed germination before winter. Abscisic acid's effects are degraded within plant tissues during cold temperatures or by its removal by water washing in out of the tissues, releasing the seeds and buds from dormancy.

In plants under water stress, ABA plays a role in closing the stomata. Soon after plants are water-stressed and the roots are deficient in water, a signal moves up to the leaves, causing the formation of ABA precursors there, which then move to the roots. The roots then release ABA, which is translocated to the foliage through the vascular system and modulates the potassium and sodium uptake within the guard cells, which then lose turgidity, closing the stomata. ABA exists in all parts of the plant and its concentration within any tissue seems to mediate its effects and function as a hormone; its degradation, or more properly catabolism, within the plant affects metabolic reactions and cellular growth and production of other hormones. Plants start life as a seed with high ABA levels. Just before the seed germinates, ABA levels decrease; during germination and early growth of the seedling, ABA levels decrease even more. As plants begin to produce shoots with fully functional leaves, ABA levels begin to increase, slowing down cellular growth in more "mature" areas of the plant. Stress from water or predation affects ABA production and catabolism rates, mediating another cascade of effects that trigger specific responses from targeted cells. Scientists are still piecing together the complex interactions and effects of this and other phytohormones.

## Auxins

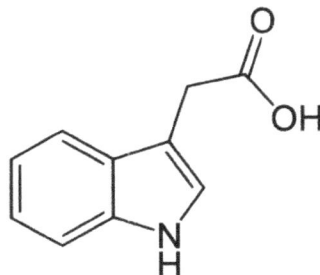

The auxin indole-3-acetic acid

Auxins are compounds that positively influence cell enlargement, bud formation and root initiation.

They also promote the production of other hormones and in conjunction with cytokinins, they control the growth of stems, roots, and fruits, and convert stems into flowers. Auxins were the first class of growth regulators discovered. They affect cell elongation by altering cell wall plasticity. They stimulate cambium, a subtype of meristem cells, to divide and in stems cause secondary xylem to differentiate. Auxins act to inhibit the growth of buds lower down the stems (apical dominance), and also to promote lateral and adventitious root development and growth. Leaf abscission is initiated by the growing point of a plant ceasing to produce auxins. Auxins in seeds regulate specific protein synthesis, as they develop within the flower after pollination, causing the flower to develop a fruit to contain the developing seeds. Auxins are toxic to plants in large concentrations; they are most toxic to dicots and less so to monocots. Because of this property, synthetic auxin herbicides including 2,4-D and 2,4,5-T have been developed and used for weed control. Auxins, especially 1-Naphthaleneacetic acid (NAA) and Indole-3-butyric acid (IBA), are also commonly applied to stimulate root growth when taking cuttings of plants. The most common auxin found in plants is indole-3-acetic acid or IAA. The correlation of auxins and cytokinins in the plants is a constant (A/C = const.).

## Cytokinins

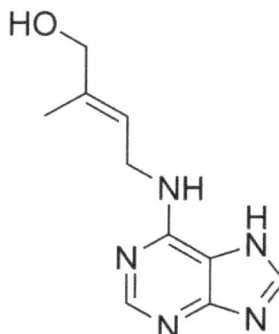

The cytokinin zeatin, the name is derived from *Zea*, in which it was first discovered in immature kernels.

Cytokinins or CKs are a group of chemicals that influence cell division and shoot formation. They were called kinins in the past when the first cytokinins were isolated from yeast cells. They also help delay senescence of tissues, are responsible for mediating auxin transport throughout the plant, and affect internodal length and leaf growth. They have a highly synergistic effect in concert with auxins, and the ratios of these two groups of plant hormones affect most major growth periods during a plant's lifetime. Cytokinins counter the apical dominance induced by auxins; they in conjunction with ethylene promote abscission of leaves, flower parts, and fruits. The correlation of auxins and cytokinins in the plants is a constant (A/C = const.).

## Ethylene

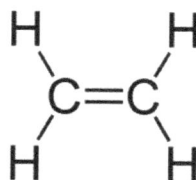

Ethylene

Ethylene is a gas that forms through the breakdown of methionine, which is in all cells. Ethylene has very limited solubility in water and does not accumulate within the cell but diffuses out of the cell and escapes out of the plant. Its effectiveness as a plant hormone is dependent on its rate of production versus its rate of escaping into the atmosphere. Ethylene is produced at a faster rate in rapidly growing and dividing cells, especially in darkness. New growth and newly germinated seedlings produce more ethylene than can escape the plant, which leads to elevated amounts of ethylene, inhibiting leaf expansion. As the new shoot is exposed to light, reactions by phytochrome in the plant's cells produce a signal for ethylene production to decrease, allowing leaf expansion. Ethylene affects cell growth and cell shape; when a growing shoot hits an obstacle while underground, ethylene production greatly increases, preventing cell elongation and causing the stem to swell. The resulting thicker stem can exert more pressure against the object impeding its path to the surface. If the shoot does not reach the surface and the ethylene stimulus becomes prolonged, it affects the stem's natural geotropic response, which is to grow upright, allowing it to grow around an object. Studies seem to indicate that ethylene affects stem diameter and height: When stems of trees are subjected to wind, causing lateral stress, greater ethylene production occurs, resulting in thicker, more sturdy tree trunks and branches. Ethylene affects fruit-ripening: Normally, when the seeds are mature, ethylene production increases and builds-up within the fruit, resulting in a climacteric event just before seed dispersal. The nuclear protein Ethylene Insensitive2 (EIN2) is regulated by ethylene production, and, in turn, regulates other hormones including ABA and stress hormones.

## Gibberellins

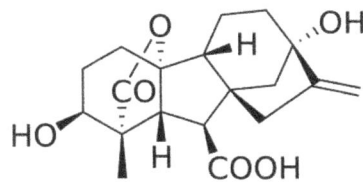

Gibberellin A1

Main function: initiate mobilization of storage materials in seeds during germination, cause elongation of stems, stimulate bolting in biennials stimulate pollen tube growth.

Gibberellins, or GAs, include a large range of chemicals that are produced naturally within plants and by fungi. They were first discovered when Japanese researchers, including Eiichi Kurosawa, noticed a chemical produced by a fungus called *Gibberella fujikuroi* that produced abnormal growth in rice plants. Gibberellins are important in seed germination, affecting enzyme production that mobilizes food production used for growth of new cells. This is done by modulating chromosomal transcription. In grain (rice, wheat, corn, etc.) seeds, a layer of cells called the aleurone layer wraps around the endosperm tissue. Absorption of water by the seed causes production of GA. The GA is transported to the aleurone layer, which responds by producing enzymes that break down stored food reserves within the endosperm, which are utilized by the growing seedling. GAs produce bolting of rosette-forming plants, increasing internodal length. They promote flowering, cellular division, and in seeds growth after germination. Gibberellins also reverse the inhibition of shoot growth and dormancy induced by ABA.

## Other Known Hormones

Other identified plant growth regulators include:

- Brassinosteroids - are a class of polyhydroxysteroids, a group of plant growth regulators. Brassinosteroids have been recognized as a sixth class of plant hormones, which stimulate cell elongation and division, gravitropism, resistance to stress, and xylem differentiation. They inhibit root growth and leaf abscission. Brassinolide was the first identified brassinosteroid and was isolated from extracts of rapeseed (*Brassica napus*) pollen in 1979.

- Salicylic acid — activates genes in some plants that produce chemicals that aid in the defense against pathogenic invaders.

- Jasmonates — are produced from fatty acids and seem to promote the production of defense proteins that are used to fend off invading organisms. They are believed to also have a role in seed germination, and affect the storage of protein in seeds, and seem to affect root growth.

- Plant peptide hormones — encompasses all small secreted peptides that are involved in cell-to-cell signaling. These small peptide hormones play crucial roles in plant growth and development, including defense mechanisms, the control of cell division and expansion, and pollen self-incompatibility.

- Polyamines — are strongly basic molecules with low molecular weight that have been found in all organisms studied thus far. They are essential for plant growth and development and affect the process of mitosis and meiosis.

- Nitric oxide (NO) — serves as signal in hormonal and defense responses (e.g. stomatal closure, root development, germination, nitrogen fixation, cell death, stress response). NO can be produced by a yet undefined NO synthase, a special type of nitrite reductase, nitrate reductase, mitochondrial cytochrome c oxidase or non enzymatic processes and regulate plant cell organelle functions (e.g. ATP synthesis in chloroplasts and mitochondria).

- Strigolactones - implicated in the inhibition of shoot branching.

- Karrikins - not plant hormones because they are not made by plants, but are a group of plant growth regulators found in the smoke of burning plant material that have the ability to stimulate the germination of seeds

- Triacontanol - a fatty alcohol that acts as a growth stimulant, especially initiating new basal breaks in the rose family. It is found in alfalfa (lucerne), bee's wax, and some waxy leave cuticles.

## Potential Medical Applications

Plant stress hormones activate cellular responses, including cell death, to diverse stress situations in plants. Researchers have found that some plant stress hormones share the ability to adversely affect human cancer cells. For example, sodium salicylate has been found to suppress proliferation of lymphoblastic leukemia, prostate, breast, and melanoma human cancer cells. Jasmonic acid, a plant stress hormone that belongs to the jasmonate family, induced death in lymphoblastic leukemia cells. Methyl jasmonate has been found to induce cell death in a number of cancer cell lines.

## Hormones and Plant Propagation

Synthetic plant hormones or PGRs are commonly used in a number of different techniques involving plant propagation from cuttings, grafting, micropropagation, and tissue culture.

The propagation of plants by cuttings of fully developed leaves, stems, or roots is performed by gardeners utilizing auxin as a rooting compound applied to the cut surface; the auxins are taken into the plant and promote root initiation. In grafting, auxin promotes callus tissue formation, which joins the surfaces of the graft together. In micropropagation, different PGRs are used to promote multiplication and then rooting of new plantlets. In the tissue-culturing of plant cells, PGRs are used to produce callus growth, multiplication, and rooting.

## Seed Dormancy

Plant hormones affect seed germination and dormancy by acting on different parts of the seed.

Embryo dormancy is characterized by a high ABA:GA ratio, whereas the seed has a high ABA sensitivity and low GA sensitivity. In order to release the seed from this type of dormancy and initiate seed germination, an alteration in hormone biosynthesis and degradation toward a low ABA/GA ratio, along with a decrease in ABA sensitivity and an increase in GA sensitivity, must occur.

ABA controls embryo dormancy, and GA embryo germination. Seed coat dormancy involves the mechanical restriction of the seed coat. This, along with a low embryo growth potential, effectively produces seed dormancy. GA releases this dormancy by increasing the embryo growth potential, and/or weakening the seed coat so the radical of the seedling can break through the seed coat. Different types of seed coats can be made up of living or dead cells, and both types can be influenced by hormones; those composed of living cells are acted upon after seed formation, whereas the seed coats composed of dead cells can be influenced by hormones during the formation of the seed coat. ABA affects testa or seed coat growth characteristics, including thickness, and effects the GA-mediated embryo growth potential. These conditions and effects occur during the formation of the seed, often in response to environmental conditions. Hormones also mediate endosperm dormancy: Endosperm in most seeds is composed of living tissue that can actively respond to hormones generated by the embryo. The endosperm often acts as a barrier to seed germination, playing a part in seed coat dormancy or in the germination process. Living cells respond to and also affect the ABA:GA ratio, and mediate cellular sensitivity; GA thus increases the embryo growth potential and can promote endosperm weakening. GA also affects both ABA-independent and ABA-inhibiting processes within the endosperm.

# Forest Pathology

Forest pathology is the research of both biotic and abiotic maladies affecting the health of a forest ecosystem, primarily fungal pathogens and their insect vectors. It is a subfield of forestry and plant pathology.

Forest pathology is part of the broader approach of forest protection.

## Abiotic Factors

There are a number of abiotic factors which affect the health of a forest, such as moisture issues like drought, winter-drying, waterlogging resulting from over-abundance or lack of precipitation such as hail, snow, rain.

Wind is also an important abiotic factor as windthrow (the uprooting or breaking of trees due to high winds) causes an obvious and direct loss of stability to a forest or its trees.

Often, abiotic factors and biotic factors will affect a forest at the same time. For example, if wind speed is 80 km per hour then many trees which have root rot (caused by a pathogen) are likely to be thrown. Higher wind speeds are necessary to damage healthier trees.

Fire, whether caused by humans or lightning and related abiotic factors also affect the health of forest.

The effects of man often alter a forest's predisposition to damage from both abiotic and biotic effects. For example soil properties may be altered by heavy machinery.

Other abiotic factors

- Nutrient imbalances: deficiencies, chemicals (toxic salts, herbicides, air pollutants)
- Stemflow which can concentrate dry deposits which via soil acidification can kill surrounding plants.
- Temperature

Biotic factors

- Fungi: Ascomycota, Basidiomycota and Fungi imperfecti

There is a category listing fungal tree pathogens and diseases.

- Oomycota: Phytophthora
- Bacteria
- Phytoplasmas
- Viruses

Insects

There is a category listing insect pests of temperate forests.

- Ips (genus) bark beetles
- Bark beetle
- Ambrosia beetle
- Cerambycidae
- Black arches

Some of these factors act in concert (all do to a degree). For example Amylostereum areolatum which is spread by the Sirex woodwasp. The fungus gains access to new trees to live off and the woodwasp larvae gain food.

## Parasitic Flowering Plants

- Many plants can parasitize trees via root to root contact. Many of these parasitic plants originate in the tropical and subtropical climates.

## Animals

Nematodes, insects especially bark beetles, mammals may browse. Browsing can be prevented with tree shelters

Part of forest pathology is forest entomology. Forest entomology includes the study of all insects and arthropods, such as mites, centipedes and millipedes, which live in and interact in forest ecosystems. Forest entomology also includes the management of insect pests that cause the degrading, defoliation, crown die-back or death of trees.

Thus the scope is wide and includes:

- Documentation of all insect species and related arthropoda in natural and man-made forests, and the study and ecology of those species.

- Description and assessment of damage to tree structures (parts of a tree), to forest stands, landscape effects and to wood products, timber in service and other ecosystem services.

- Eradication of recently introduced pests, or long-term management of established exotics and indigenous pests, to minimise losses in wood quality and wood production, and to reduce tree mortality.

- Assessments of forest operations, or of management impacts, on the invertebrate fauna, and the alleviation of any adverse effects on these invertebrates.

## Hazard Trees

- The likelihood of property damage or personal injury due to tree failure. Hazard includes not only the tree's condition, but the potential target as well. Rating systems, procedures and guidelines have been developed for decision making but knowledge, judgement, and experience are an important part to the process.

## Pathogens That Affect Trees

There is a category listing tree diseases.

- Armillaria sp which causes white rot root disease

- Cenangium

- Hymenoscyphus fraxineus which causes Ash Dieback

- Heterobasidion annosum which causes Annosum or red root rot, the economically most significant pathogen in the Northern hemisphere.

- Chestnut blight

- Rickettsia which causes possibly this citrus greening disease

- Spiroplasma

- Dutch elm disease

- Ink disease

- Emerald ash borer

- Olive tree pathology

- Witch's broom

- White Pine Blister Rust

- Phytophthora cinnamomi which causes root rot

- Phytophthora ramorum which causes sudden oak death

- Polypore or bracket fungus

- Tinder conk

## Signs and Symptoms

Symptoms are a result of a pathogen:

- Blight

- Burl

- Canker

- Chlorosis

- Drunken trees

- Forest dieback

- Gall

- Girdling

- Leaf scorch

- Root rot

- Virescence

- Wilt disease

Signs are the visible presence of a part of a pathogen:

- Ascus is a part of an acomycota fungus.

- Conk (fungi) is the fruiting body of a bracket fungus.

- Hypha are collectively called a mycelium

- Mycelial cord or rhizomorphs

## Pathology Detection

This can be done by dogs or machines smelling the trees. It is similar to noses used to find truffles. It can also be done by monitoring and identification can happen via tree clinics, experts such as arborists or even non-experts through citizen science.

It is important to consider the disease triangle when evaluating pathologies. Demonstration of suspected active agents can be done by confirmation of Koch's postulates.

## References

- Srivastava, L. M. (2002). Plant growth and development: hormones and environment. Academic Press. p. 140. ISBN 0-12-660570-X.

- Öpik, Helgi; Rolfe, Stephen A.; Willis, Arthur John; Street, Herbert Edward (2005). The physiology of flowering plants (4th ed.). Cambridge University Press. p. 191. ISBN 978-0-521-66251-2.

- Weier, Thomas Elliot; Rost, Thomas L.; Weier, T. Elliot (1979). Botany: a brief introduction to plant biology. New York: Wiley. pp. 155–170. ISBN 0-471-02114-8.

- Osborne, Daphné J.; McManus, Michael T. (2005). Hormones, signals and target cells in plant development. Cambridge University Press. p. 158. ISBN 978-0-521-33076-3.

- Roszer T (2012) Nitric Oxide Synthesis in the Chloroplast. in: Roszer T. The Biology of Subcellular Nitric Oxide. Springer New York, London, Heidelberg. ISBN 978-94-007-2818-9

# Permissions

# Index